Cosmic Origins

Bridging the Big Bang and Brahman

Dr. Manmohan Chaturvedi

Dedication

This book is dedicated to the giant minds both in empirical science and in spirituality since beginning of recorded history who have helped us get a coherent view of the external world and the world within, thus facilitating a holistic view of our reality

Contents

Foreword

1

Foreword

by

Professor Valentina Kazaryan, Doctor of Science, M.V. Lomonosov

Moscow State University (Philosophy Perspective)

and

Anna Sidorova-Biryukova, Ph.D., Russian Quantum Center,

Moscow (Scientific Perspective)

Science is approaching the point where its cognitive instruments may be exhausted. In order to make further progress, science must turn to other areas of human knowledge. Once in the 1980s, the Dalai Lama said to Carl von Weizsäcker that "Buddhism has much to learn from physics," and that was true, but in fact, science has also much to learn from spiritual disciplines.

Physics currently finds itself in an awkward state: almost everything that could be verified in experiment has already been verified, and most theoretical predictions that could yield something new are practically out of reach of the means at our disposal. A way out of this impasse is to dismiss the rule "shut up and count" and to let cultural and spiritual knowledge have a higher impact on science. The effect may be beneficial, but we should be ready to meet deep transformations in the very foundations of science with an open mind.

This book, 'Cosmic Origins: Bridging the Big Bang and Brahman' of Manmohan Chaturvedi captures the described tendency of science to revise the foundations and search for completely new paths and means to understand reality. Unexpectedly, oriental religious philosophies, especially Hinduism, turn out to be close to modern physics in many aspects. This fact was already noticed by the authors of the quantum theory Heisenberg, Oppenheimer, Schrödinger. The goal of science now is to listen to the wisdom of centuries, to hear its hints and warnings, to find a source of inspiration there and, if lucky, to make a new step forward.

This path is evidently not easy to go. It is difficult for physics to learn from spiritual teaching, and it is difficult for the latter to learn from physics, since they are focused on different subject areas and use different methods. Even when they pronounce the same words, they say different things, because the meaning of words depends on the context that sets the interpretation. Perhaps, here the Bohr's complementarity principle works, when different observers unify their views on the problems or objects that seem similar from their positions. So far, this is the only natural way to synthesize different approaches.

Another difficulty for cross-learning is increasingly narrow specialization of the branches of science. It is extremely difficult to pave parallel paths between different well-done radial trenches of knowledge. We see the recently increasing cross-interest of rather distant disciplines, for example, quantum physics and neuroscience, however, so far, the scientific community still feels a shortage of works similar to the one presented by Manmohan Chaturvedi.

The ideas discussed in the book are non-trivial and rather complicated, however, the book is intended for a wide range of thinking readers. The author strives to avoid special terms and to give clear explanations which can be understood by a reader who is non-expert in physics or Vedanta but is ready for serious intellectual work.

We hope that the unusual book by Manmohan Chaturvedi will arouse the deserved interest of the audience, attract fresh minds to the development of the concept of "parallels" and contribute to the progress of this promising direction of thought.

Professor Valentina Kazaryan, Doctor of Science, M.V. Lomonosov

Moscow State University

Link : https://philos.msu.ru/node/346

and

Anna Sidorova-Biryukova, Ph.D., Russian Quantum Center

Moscow

Link: https://www.rqc.ru/

2

Forward

by

Swami Nikhilananda Saraswati (Vedantic Perspective)

Since the dawn of consciousness, humanity has grappled with fundamental questions about our existence and the nature of the universe. Where did we come from? How did the cosmos begin? What forces shape the reality we perceive? These inquiries have driven the development of both scientific investigation and spiritual contemplation throughout human history.

In our modern era, we find ourselves at a unique juncture where two seemingly disparate paths of understanding converge in their pursuit of cosmic truth: the scientific theory of the Big Bang and the Hindu philosophical concept of Brahman. While these approaches may appear fundamentally different at first glance, a closer examination reveals intriguing parallels and complementary insights that can enrich our understanding of the universe and our place within it.

This book, 'Cosmic Origins: Bridging the Big Bang and Brahman' aims to explore the common ground between these two perspectives, bridging the perceived gap between scientific empiricism and spiritual intuition. By doing so, we hope to offer readers a more holistic and nuanced view of cosmic origins and the nature of reality itself.

This book is not an attempt to force a superficial unity where none exists. Rather, it is an exploration of genuine points of convergence and complementarity, as well as an honest examination of differences and challenges. It is an invitation to expand our understanding, to hold multiple perspectives simultaneously, and to see how different approaches to knowledge can inform and enrich each other.

The author of this book, Dr Chaturvedi, has keen interest in both modern Cosmology research (An Engineer and Researcher by training) and Spirituality (having been initiated formally in Meditation process in year 1982). This unique combination may add depth to the discourse while making the reading an interesting experience to the reader.

I wish this book a great success.

Swami Nikhilananda Saraswati*

*Swami Nikhilananda Saraswati, is a Student and a Teacher of Advaita Vedanta. He is a disciple of Swami Chinmayananda Saraswati and Swami Paramananda Bharati. Link to his you tube is given below:

https://www.youtube.com/@swami.nikhilananda.saraswati

About the Author

Dr Manmohan Chaturvedi is an engineer by profession and holds a PhD. from IIT Delhi. He served the Indian Air Force for about 35 years and retired from the rank of Air Commodore.

He has published following 4 books:

a) Conquering Fear of Death by Listening to Whispers within
b) Explore within Self for Meaning of Life
c) Cybersecurity from Management Perspective
d) Global Trends in Digital Technologies- Digital Policies and Regulations

He had interest in Spirituality since college days. With a purpose to share understanding gathered while reading scriptures and books on philosophy, he is regularly sharing his insights in a Spiritual blog.

Link to the blog 'Search within' is shared below:

https://search-within-self.blogspot.com/

Preface

"I have tried to read philosophers of all ages and have found many illuminating ideas but no steady progress toward deeper knowledge and understanding. Science, however, gives me the feeling of steady progress: I am convinced that theoretical physics is actual philosophy."

- Max Born (German-British physicist and mathematician)

"... with all our science, we are not a step closer to understanding the essence than an old Indian sage."

- K. Jaspers (German-Swiss psychiatrist and philosopher)

'Above two epigraphs at first glance seem contradictory to each other; however, this contradiction could be better understood as complementarity, when two observers look at the same phenomenon from mutually orthogonal directions.

One of the authors is a physicist, the other is a philosopher, and the subject they speak about can be called "metaphysics", or the science of eternal questions. The philosopher emphasizes their eternal character, i.e. the impossibility for a human mind to reach the finish line, while the physicist talks about the possibility and even the human duty to go along this endless path.

This reminds of how Alice in the Looking-Glass Land had to run as fast as she could just to stay in place.'[1]

This brings to light the challenge that is inherent in viewing the world through the two lenses of Science and Spirituality.

The idea of a book on this theme occurred to me few months ago while reading a book title 'Introducing Stephen Hawking: A Graphic Guide' by J.P. McEvoy . This guide explores the fascinating life of Stephen Hawking, and the evolution of his work from his days as a student. While describing scientific view of the origin of our universe, concept of 'Singularity' was described. A 'singularity' is a region of space where the laws of physics break down and the curvature of space-time becomes infinite. Possibly, physicists are still attempting a coherent theory to describe events that happened before big bang and birth of our universe. Big bang as a phenomenon seems to be on sound theoretical ground because of the supporting evidence observed by science.

This idea that a 'singularity' is a point where known laws of physics breakdown triggered in my mind something similar described in 'Patanjali Yoga Sutra' in Hindu scriptures for exploring spiritual aspect of our personality . The state of 'Samadhi' (total absorption) reached by a Yogi (a practitioner of the discipline of Spirituality) during meditation is a state that the Yogi is unable to describe using known methods to articulate one's experience. As per Patanjali Yoga Sutra, mind/intellect, our tool kit for recording experience, is suspended in this event of total self-absorption. However, the Samadhi state is **not** absence

---

[1] *Sidorova-Biryukova, A. (2020). Theoretical Physics and Indian Philosophy: Conceptual Coherence. arXiv preprint arXiv:2004.02150.*

of any experience. As per the Hindu scriptures, this state is experienced by many Spiritual seekers and is known to transform the world view of the seeker.

I have been practicing meditation since 1982 and though I cannot claim to have achieved a transformation of my personality as described by Patanjali, I seem to grasp this phenomenon in an intuitive manner.

I started speculating if this transition to 'Samadhi' state could be considered a 'Singularity' in our mental process similar to that described for physical world at the birth of our universe.

My search for these parallels took me to many books and speculations written by eminent physicists and philosophers. 'Tao Of The Physics' by Fritjo Capra, a physicist by training caught my attention. The 'Tao of Physics' is a classic study of the prominent similarities between Eastern mysticism and modern physics. The central premise of the book is that the ancient and mystical traditions of the East hold a rational theoretical framewoek, which can accommodate the advanced Western theories of the modern world.

However, I was cautious. I am a technocrat by profession. Spirituality is an add on from my college days. My MTech from IIT Delhi in 1985 (around the same time when I took formal initiation into meditation practice in 1982) in Optical Communication and Opto-Electronics gave me a chance to get some rudimentary understanding about wave-particle duality of light and quantum physics

It seems the seeds of duality of the physical world were sown in those years and though my life as an Air Force officer did not offer me much choice to distant my self from Newtonian laws of

Physics (Aeroplanes fly using physical laws !) , somewhere deep within I was envisaging parallels in the two domains of Science and Spirituality.

Any search for truth can possibly take two different approaches. Deductive and Inductive methodologies are two fundamental approaches to research and reasoning. Both approaches have their use and a holistic approach may benefit by suitable combination of the two.

Modern science tends to take Deductive approach using empirical observations and in its search for ultimate cause of the universe has reached the quantum particles and faces limits of uncertainty in ongoing search.

Vedantic approaches postulating Brahman as the substratum of our observed world takes an Inductive approach, where search for the ultimate cause entails an inward journey through a well-defined practice by the seeker of truth. The truth is realised through a subjective experience and is difficult to describe since our mind and intellect are left behind during the final phase. This characteristic of the spiritual insight is not acceptable to modern science where empirical evidence shareable with other researchers is considered essential to a valid research.

Certain limitations faced by modern research by way of quantum entanglement and observer's role in quantum measurements are areas of focus by modern physics.

This book, "Cosmic Origins: Bridging the Big Bang and Brahman," emerges from a conviction that there is immense value in bringing these seemingly disparate perspectives into dialogue. It is an attempt to weave together the empirical discoveries of modern science with the profound insights of Hindu philosophy, particularly the concept of Brahman, to create a more comprehensive understanding of our cosmic origins and the nature of reality.

The Big Bang theory, with its description of the universe emerging from an infinitesimally small, infinitely dense point, resonates in intriguing ways with the Hindu concept of the universe cyclically manifesting from and dissolving back into Brahman.

Quantum mechanics, with its emphasis on the role of the observer and the interconnectedness of all things, echoes ancient non-dual philosophies. The scientific concept of emergence, where complex systems arise from simple rules, parallels spiritual ideas about the manifestation of diversity from a fundamental unity.

Yet, I also became acutely aware of the challenges in bringing these perspectives together. The rigorous empiricism of science and the intuitive insights of spiritual traditions seem incompatible at first glance. There are real and important differences in methodology, epistemology, and worldview that cannot be glossed over or ignored.

Throughout the chapters that follow, we will journey from the earliest moments of cosmic history to the far reaches of human understanding. We will explore the latest findings in cosmology and quantum physics alongside profound concepts from Hindu philosophy. We will grapple with questions of consciousness, time, the nature of reality, and the place of humanity in the cosmos.

Our goal is not to arrive at final answers, but to open up new ways of thinking about these fundamental questions. We aim to foster a dialogue between science and spirituality that respects the integrity of both while exploring the rich territory where they intersect.

This book is intended for a wide audience – scientists curious about the philosophical implications of their work, spiritual seekers interested in how modern science might inform their understanding, and general readers fascinated by the big questions of existence. No specialized knowledge of physics or Hindu philosophy is assumed, though a spirit of open-minded inquiry and a willingness to grapple with complex ideas will greatly enhance the reading experience.

As we embark on this journey of exploration and synthesis, I invite you to approach these ideas with both critical thinking and open-hearted wonder. Question assumptions, examine evidence, and be willing to consider new perspectives. At the same time, allow yourself to be awed by the mystery and majesty of the cosmos, to feel your connection to the vast processes that have shaped our universe.

It is my hope that this book will not only expand your understanding of cosmic origins but also deepen your sense of connection to the universe and your place within it. Perhaps, in bridging the Big Bang and Brahman, we can come to a more holistic vision of reality – one that honours both the rigorous findings of science and the profound insights of spiritual wisdom.

As you turn these pages, may you feel the thrill of cosmic exploration and the depth of philosophical contemplation. May this journey through the origins of the cosmos also be a journey of self-discovery. In seeking to understand our cosmic origins, we are ultimately seeking to understand ourselves and our place in this vast, mysterious, and awe-inspiring universe.

This book has **12 main chapters** and **7 appendices** in addition to prologue and epilogue. Use of these 7 appendices may help in providing useful information on specific topics of interest to the readers while keeping the main chapters concise and easy to follow. In **bibliography** section alphabetically relevant references are listed for a deep dive by interested reader. The **index section** provides alphabetic index of important topics with page numbers. Using index section reader may look for topics of his interest with ease. Based on feedback received from readers certain parts of the chapters are rewriiten to improve readability and clarify some concepts using easy to understand description.

In compiling this book, I have extensively used internet resources and AI tools to help me organize the text on various facets of current research in cosmology and Hindu texts on spirituality. Thus an attempt is made to ensure that salient and topical texts in both domains of our interest are adequately covered. My experience and insights helped me to validate and improve the contents further.

Welcome to the adventure of "Cosmic Origins: Bridging the Big Bang and Brahman." Let us begin our exploration of the greatest mystery of all – the nature of existence itself.

Manmohan Chaturvedi

Prologue

Imagine, for a moment, that you could zoom out from your current position, expanding your view until you could see the entire Earth, a blue marble suspended in the vastness of space. Keep zooming out, past the moon, beyond the solar system, until our galaxy becomes just one point of light among billions. Further still, until you can see the vast cosmic web of galaxies stretching across the observable universe.

Now, reverse direction. Zoom in, past galaxies and solar systems, down to the Earth, to a single human being. Keep going, past cells and molecules, down to the level of atoms, then to subatomic particles, and finally to the quantum foam of spacetime itself, where reality becomes a blur of probability and potential.

In this journey across scales, from the unimaginably vast to the infinitesimally small, we traverse the full scope of our current understanding of the cosmos. Yet, remarkably, we find that at both extremes – the cosmic and the quantum – our usual notions of reality begin to break down. The solid, predictable world of our everyday experience gives way to counter-intuitive realms that challenge our deepest assumptions about the nature of existence.

It is in these boundary realms, where our understanding falters, that science and spirituality often find common ground. Both reach for languages and concepts to describe realities that defy ordinary perception and logic. Both confront us with mysteries that awaken a sense of awe and wonder.

Consider, for a moment, two descriptions of the origin of the universe:

In the first, the entire cosmos explodes into being from an infinitesimally small point, unimaginably hot and dense. Space itself expands faster than the speed of light, cooling as it grows, allowing fundamental particles to condense out of pure energy. Over billions of years, these particles combine to form atoms, stars, galaxies, and eventually, on at least one small planet, life capable of contemplating its own origins emerges.

In the second, the universe emanates from an eternal, unchanging reality beyond time and space. This ultimate reality, through its own creative power, manifests as the myriad forms and phenomena we perceive. The entire cosmos is seen as a dynamic play of consciousness, cyclically manifesting and dissolving like waves on an infinite ocean.

The first description might be recognized as a simplified account of the Big Bang theory, the cornerstone of modern cosmology. The second resonates with Hindu philosophical concepts, particularly the idea of Brahman as the source and substance of all existence.

At first glance, these two accounts might seem irreconcilable. One speaks the language of physics – of particles, energy, and expanding space. The other uses the vocabulary of metaphysics – of ultimate reality, consciousness, and cycles of manifestation. Yet, as we'll explore in this book, there are intriguing parallels and points of convergence between these perspectives.

Both, for instance, suggest that the universe as we know it had a beginning. Both imply a progression from simplicity to complexity. Both grapple with concepts that strain the limits of human understanding – be it the singularity of the Big Bang or the difficult to describe ineffable nature of Brahman.

More profoundly, both scientific cosmology and Hindu philosophy lead us to question our ordinary perception of reality. Science reveals a universe far stranger and more mysterious than our everyday experience suggests – a cosmos where time can bend, where particles can be entangled across vast distances, where most of what exists is invisible to us. Hindu thought, particularly in its non-dualistic schools, challenges our notion of a solid, independent reality, suggesting that the world of appearances is a kind of illusion obscuring a deeper, unified truth.

As we stand here in the early 21st century, we find ourselves at a unique juncture in human history. We have peered back to the earliest moments of cosmic history and plumbed the depths of the subatomic world. We have begun to grasp the vast scales of space and time that define our universe. At the same time, we have inherited millennia of philosophical and spiritual insight into the nature of reality and consciousness.

The question before us now is: Can we integrate these different ways of knowing? Can we create a more comprehensive understanding of our cosmic origins and the nature of reality itself?

1. Introduction – Two Paths to Understanding Creation

1. The Importance of Dialogue between two doctrines

As described earlier in preface to the book, we seem to have two contesting approaches to unravel the mystery of our universe. At first glance, the scientific model of the Big Bang and the Hindu concept of Brahman might seem irreconcilable. One is based on empirical observation and mathematical models, while the other stems from philosophical reflection and spiritual insight. However, as we delve deeper into both perspectives, we find surprising areas of convergence and complementarity.

This dialogue between science and spirituality is not merely an academic exercise. It has profound implications for how we understand our place in the cosmos and how we approach the big questions of existence. By exploring the common ground between these two approaches, we can:

1. Gain a more comprehensive understanding of reality

2. Bridge the perceived divide between scientific and spiritual worldviews

3. Foster greater cross-cultural and interdisciplinary dialogue

4. Develop new ways of thinking about cosmic origins and evolution

Moreover, this exploration can help us recognize the limitations of both scientific and spiritual approaches when taken in isolation. Science, for all its empirical rigor, cannot yet explain what came "before" the Big Bang. The question "why is there something rather than nothing" is a fundamental metaphysical question in physics that may not have an absolute answer.

Similarly, spiritual insights, while profound, often lack the predictive power and testability of scientific theories.

By bringing these perspectives into conversation, we can create a richer, more nuanced understanding of the cosmos and our place within it.

2. Key Themes of the Book

As we embark on this journey of exploration and synthesis, several key themes will guide our inquiry:

2.1 The Nature of Time and Eternity

Both modern physics and Hindu philosophy challenge our conventional notions of time. Einstein's theory of relativity reveals that time is not absolute but relative, warped by gravity and motion. In Hindu thought, time is often seen as cyclical rather than linear, with universes going through vast cycles of creation and dissolution. However, In Vedanta time is mentioned of two types. One that which can be counted like seconds days years etc. and another which is absolute time identical with Brahman. In Bhagavad Gita these two are referred as "kala kalayatam aham" (Translated as "Among subduers I am time")in chapter 10 and as akshaya kala (eternal, immortal, and indestructible time) in chapter 11. In the Bhagavad Gita, the word kāla comes from kalayati, which means "to take count of". Time counts and controls the lifespan of all beings, and it buries all events in nature.

We will explore how these different conceptions of time relate to our understanding of cosmic origins and the nature of reality itself.

2.2 Consciousness and the Cosmos

The role of consciousness in the universe is a point of fascinating convergence between scientific and spiritual perspectives. In quantum physics, the act of observation seems to play a crucial role in determining reality. Similarly, in Advaita Vedanta, a prominent school of Hindu philosophy, consciousness is seen as the fundamental basis of all existence.

We will examine theories of panpsychism, the hard problem of consciousness, and the Hindu concept of Atman (the individual self) and its relationship to Brahman.

2.3 The Limits of Knowledge

Both science and spirituality recognize inherent limits to human knowledge. In science, these limits are expressed through principles like Heisenberg's uncertainty principle and Gödel's incompleteness theorems. In Hindu philosophy, the concept of Maya (illusion) suggests that our ordinary perception of reality is limited and, in some sense, illusory.However, the world of appearance too is fundamentally identical with Brahman. Just as the individual self is fundamentally identical with Brahman.

We will explore these epistemological limits and their implications for our understanding of cosmic origins.

2.4 Unity and Diversity

The scientific quest for a "Theory of Everything" that unifies all fundamental forces parallels the Hindu conception of underlying unity expressed through apparent diversity. We will examine how these ideas relate and what they suggest about the fundamental nature of reality.

2.5 Ethical Implications

Our understanding of cosmic origins and the nature of reality has profound implications for how we live our lives and treat our world. We will explore the ethical dimensions of both scientific cosmology and Hindu philosophy, considering how these worldviews can inform our approach to environmental stewardship, social justice, and personal growth.

3. A Note on Approach

As we navigate the complex territory where science and spirituality intersect, it's crucial to maintain a balanced and respectful approach. This book does not aim to "prove" one perspective right and the other wrong, nor does it seek to forcibly reconcile fundamentally different ways of knowing.

Instead, our goal is to:

1. Present both scientific and Hindu perspectives accurately and fairly

2. Explore areas of convergence and divergence with an open mind

3. Encourage critical thinking and personal reflection

4. Foster a spirit of inquiry and dialogue

We recognize that both science and spirituality are vast fields with diverse viewpoints. When discussing the "scientific perspective," we will focus primarily on mainstream, peer-reviewed research in cosmology and related fields. Our exploration of Hindu philosophy will draw mainly from classical texts and respected contemporary interpretations, acknowledging the diversity of thought within the Hindu tradition.

It's also important to note that while this book focuses on the Big Bang theory and the concept of Brahman, these are not the only perspectives on cosmic origins. Many other scientific theories, religious traditions, and philosophical systems offer valuable insights into the nature of the universe and our place within it. While we can't explore all of these in depth, we will occasionally reference other viewpoints to provide context and contrast.

4. The Journey Ahead

In the chapters that follow, we will delve deeper into the specifics of Big Bang cosmology and the concept of Brahman, exploring their historical development, key principles, and contemporary interpretations. We will then examine various aspects of cosmic origins and the nature of reality through both lenses, seeking points of convergence and complementarity.

Our journey will take us from the earliest moments of cosmic history to the far future of the universe, from the smallest quantum fluctuations to the largest structures in the cosmos. Along the way, we will grapple with profound questions about the nature of time, space, consciousness, and reality itself.

As we stand at the intersection of empirical observation and spiritual insight, we have a unique opportunity to enrich our understanding of the cosmos and our place within it. By bridging the Big Bang and Brahman, we may discover new ways of thinking about the greatest mysteries of existence and forge a more holistic vision of our cosmic heritage.

2. The Big Bang – Modern Cosmology's Origin Story

1. Introduction: The Birth of Modern Cosmology

The 20th century witnessed a revolution in our understanding of the universe. At the dawn of the 1900s, most scientists believed the universe was static and eternal. By the century's end, a radically different picture had emerged: a cosmos with a definite beginning, expanding and evolving over billions of years. This transformation in our cosmic perspective is largely due to the development and refinement of the Big Bang theory.

The Big Bang theory stands as one of the most significant scientific achievements of the modern era. It provides a comprehensive framework for understanding the origin and evolution of the universe, supported by a wealth of observational evidence. In this chapter, we will explore the history, key concepts, and evidence supporting the Big Bang theory, as well as some of its limitations and ongoing areas of research.

2. Historical Development of the Big Bang Theory

2.1 Early 20th Century: A Static Universe

At the turn of the 20th century, the prevailing view among scientists was that the universe was static and eternal. This perspective was so entrenched that when Albert Einstein developed his general theory of relativity in 1915, he introduced a "cosmological constant" to ensure his equations described a static universe, even though his original formulation suggested an expanding or contracting cosmos.

2.2 Edwin Hubble and the Expanding Universe

The first major challenge to the static universe model came from the American astronomer Edwin Hubble. In the 1920s, Hubble made two groundbreaking discoveries:

1. He proved that the nebulae observed in the night sky were actually distant galaxies, vastly expanding the known size of the universe.

2. He discovered that these galaxies were moving away from us, and the farther away they were, the faster they were receding. This relationship, now known as Hubble's Law, provided the first empirical evidence for an expanding universe.

Hubble's observations forced a radical rethinking of cosmic history. If the universe is expanding, then logically, it must have been smaller in the past. Tracing this expansion backwards in time leads to the conclusion that the universe had a beginning.

2.3 Georges Lemaître and the "Primeval Atom"

While Hubble was making his observations, a Belgian priest and physicist named Georges Lemaître was independently developing theoretical models of an expanding universe. In 1927, Lemaître proposed what he called the "hypothesis of the primeval atom," suggesting that the universe began as an incredibly dense, hot singularity that he poetically described as a "cosmic egg exploding at the moment of creation."

Lemaître's ideas were initially met with skepticism, including from Einstein himself. However, as observational evidence mounted, his model gained traction in the scientific community.

2.4 George Gamow and the Hot Early Universe

In the 1940s, physicist George Gamow and his collaborators, including Ralph Alpher and Robert Herman, further developed Lemaître's ideas. They proposed that the early universe was extremely hot and dense, dominated by radiation. As the universe expanded and cooled, it would have gone through a series of phase transitions, eventually leading to the formation of atoms, stars, and galaxies.

Crucially, Gamow's team predicted that if their model was correct, there should be a faint afterglow of radiation permeating the universe – a relic of its hot, dense early state. This prediction would later lead to one of the most important discoveries in the history of cosmology.

2.5 Coining of the Term "Big Bang"

Interestingly, the term "Big Bang" was not initially used by proponents of the theory. It was coined in 1949 by astronomer Fred Hoyle, who was actually a staunch opponent of the idea. Hoyle used the term somewhat dismissively during a BBC radio broadcast, but it caught on and became the popular name for the theory.

3. Key Concepts of the Big Bang Theory

The Big Bang theory, as it is understood today, encompasses several key concepts:

3.1 The Singularity

The Big Bang theory posits that the universe began as an infinitesimally small, infinitely dense point known as a singularity. This concept represents the limit of our current understanding, as our physical theories break down when describing such extreme conditions.

3.2 Rapid Expansion

From this initial state, the universe underwent a period of incredible expansion. This expansion was not an explosion in the conventional sense, happening within space, but an expansion of space itself.

3.3 Cooling and Phase Transitions

As the universe expanded, it also cooled. This cooling led to a series of phase transitions, analogous to how water changes from steam to liquid to ice as it cools. These phase transitions marked significant shifts in the dominant forces and particles in the universe.

3.4 Formation of Elements

In the first few minutes after the Big Bang, conditions were right for the formation of the lightest elements: primarily hydrogen and helium, with trace amounts of lithium. This process is known as Big Bang nucleosynthesis.

3.5 Decoupling and the Cosmic Microwave Background

About 380,000 years after the Big Bang, the universe had cooled enough for electrons to combine with nuclei, forming neutral atoms. This allowed light to travel freely through space for the first time, creating the cosmic microwave background radiation that we can still detect today.

3.6 Structure Formation

Over billions of years, slight irregularities in the distribution of matter grew under the influence of gravity, eventually leading to the formation of stars, galaxies, and the large-scale structure we observe in the universe today.

4. Evidence Supporting the Big Bang Theory

The Big Bang theory is supported by a wealth of observational evidence. Here are some of the key pieces of evidence:

4.1 The Expansion of the Universe

As originally observed by Edwin Hubble and confirmed by countless observations since, galaxies are moving away from us, and the farther away they are, the faster they're receding. This is exactly what we would expect to see if the universe is expanding from an initial hot, dense state.

4.2 The Cosmic Microwave Background (CMB)

Predicted by Gamow and his colleagues and discovered accidentally by Arno Penzias and Robert Wilson in 1964, the CMB is often described as the "afterglow" of the Big Bang. This faint radiation, coming from all directions in space, has a temperature of about 2.7 Kelvin and displays tiny fluctuations that provide a wealth of information about the early universe.

The properties of the CMB match the predictions of the Big Bang theory with incredible precision. The discovery and subsequent detailed mapping of the CMB (by satellites like COBE, WMAP, and Planck) have provided some of the strongest evidence for the Big Bang model.

4.3 The Abundance of Light Elements

The Big Bang theory predicts specific ratios for the abundance of light elements in the universe, particularly hydrogen and helium. Observations of the composition of stars and interstellar gas match these predictions remarkably well, providing strong support for Big Bang nucleosynthesis.

4.4 The Age of the Universe

By measuring the expansion rate of the universe and the distances to the oldest stars, astronomers have determined that the universe is about 13.8 billion years old. This age is consistent with the Big Bang model and matches the age derived from studies of the cosmic microwave background.

4.5 Large-Scale Structure of the Universe

Computer simulations based on the Big Bang model predict a specific pattern of matter distribution on large scales, often described as a "cosmic web" of galaxies and galaxy clusters. Observations of the actual distribution of galaxies in the universe match these predictions closely.

5. Refinements and Extensions to the Big Bang Theory

While the basic framework of the Big Bang theory has remained robust, it has undergone significant refinements and extensions over the years:

5.1 Inflation Theory

Proposed by Alan Guth in 1980, inflation theory suggests that the very early universe underwent a brief period of exponential expansion. This idea helps explain several puzzling features of the universe, including its remarkable uniformity on large scales and its apparent "flatness" in terms of overall geometry.

5.2 Dark Matter

Observations of galaxy rotation curves and gravitational lensing suggest that there is far more matter in the universe than we can see.

Galaxy rotation curves plot the orbital speeds of stars and gas in a galaxy against their distance from the galaxy's center. The observed speeds don't follow the same rules as other orbital systems, such as planets and moons.

Gravitational lensing occurs when a massive celestial body, like a galaxy cluster, bends light from a distant source. The amount of bending is described by Albert Einstein's general theory of relativity.

This discrepancy in galaxy rotation curves and gravitational lensing suggest the existence of dark matter.This "dark matter" is thought to play a crucial role in the formation of cosmic structure and the dynamics of galaxies.

5.3 Dark Energy

Dark energy is a mysterious force that pushes outward to expand the universe faster and faster. It's estimated to make up 70% of the universe.

The discovery of dark energy was made in the late 1990s by two teams of astronomers who were studying Type Ia supernovas. They observed that more distant supernovas were fainter than expected, which indicated that the universe's expansion was accelerating. This discovery was a surprise because scientists previously believed that gravity would slow the expansion of the universe over time. This "dark energy," is a form of energy that permeates all of space and acts against the pull of gravity. Its origin is not understood so far.

5.4 Baryogenesis

The Big Bang should have produced equal amounts of matter and antimatter, which would have annihilated each other. The fact that we live in a matter-dominated universe suggests some mechanism created a slight imbalance in favour of matter. Understanding this process, known as baryogenesis, is an active area of research.

6. Limitations and Open Questions

While the Big Bang theory is remarkably successful, it does have limitations and leaves several important questions unanswered:

6.1 The Initial Singularity

The theory breaks down when describing the universe at the very moment of the Big Bang. Our current physical theories cannot describe the conditions of infinite density and temperature predicted at the singularity.

6.2 The Cause of the Big Bang

The Big Bang theory describes the evolution of the universe from its earliest known state, but it doesn't explain what caused this state or what, if anything, came before it.

6.3 The Nature of Dark Matter and Dark Energy

While there is strong evidence for the existence of dark matter and dark energy, we still don't know what they are. Identifying the particles that make up dark matter and understanding the nature of dark energy are major goals of modern physics and astronomy.

6.4 The Matter-Antimatter Asymmetry

As mentioned earlier, we don't fully understand why the universe contains so much more matter than antimatter.

6.5 The Ultimate Fate of the Universe

While current observations suggest that the universe will continue to expand indefinitely, our understanding of dark energy is not complete enough to make definitive predictions about the far future of the cosmos.

7. The Big Bang and Fundamental Physics

The study of the early universe provides a unique laboratory for testing theories of fundamental physics. The extreme conditions present in the first fraction of a second after the Big Bang cannot be replicated in any earthly laboratory, making cosmology an essential tool for exploring the laws of nature at their most fundamental level.

7.1 Particle Physics and the Early Universe

Particle physics, also known as high-energy physics, is the study of the fundamental particles and forces that make up matter and radiation. It involves studying the structure and forces of subatomic particles, which are the smallest detectable particles.

The first tiny fraction of a second after the Big Bang saw energies far higher than those achievable in particle accelerators. Studying this period can provide insights into particle physics beyond the Standard Model.

7.2 Quantum Gravity

At the Planck scale, approximately 10^{-43} seconds after the Big Bang, the effects of both quantum mechanics and gravity would have been significant. Understanding this era requires a theory of quantum gravity, which remains one of the holy grails of theoretical physics.

Quantum gravity is a theoretical framework that attempts to describe gravity in situations where quantum effects are important, such as near black holes or in the early stages of the universe. It's considered one of the most important unsolved problems in fundamental physics.

Quantum gravity combines general relativity and quantum mechanics to describe the microstructure of spacetime. It's expected to be relevant in extreme physical situations, such as near the Planck scale, where the fundamental constants of the theories come together to form units of mass, length, and time.

There is no universally accepted and confirmed theory of quantum gravity. Instead, there are several research lines that offer tentative solutions to the problem.

7.3 String Theory and the Multiverse

String theory is one of the theories that attempts to unify the macroscopic world of gravity with the microscopic world of quantum physics.

String theory is a concept in physics that states the universe is constructed by tiny vibrating strings, smaller than the smallest subatomic particles. As these fundamental strings twist, fold and vibrate, they create matter, energy and all sorts of phenomena like electromagnetism, gravity, etc.

Some versions of string theory, a candidate for a "theory of everything," suggest the possibility of multiple universes. While highly speculative, these ideas represent attempts to extend our understanding beyond the limits of the Big Bang theory.

8. The Big Bang and Philosophy

The Big Bang theory has profound philosophical implications. It suggests that the universe had a beginning, a concept that aligns with some religious and philosophical traditions but challenges others. It raises questions about the nature of time, the possibility of creation ex nihilo (out of nothing), and humanity's place in the cosmos.

The theory has been interpreted in various ways by different philosophical and religious traditions. Some see it as compatible with the idea of a divine creator, while others view it as supporting a naturalistic worldview. The Big Bang theory itself, however, is a scientific model that neither confirms nor denies any particular metaphysical or theological position.

9. Conclusion: The Big Bang as a Scientific Revolution

The development of the Big Bang theory represents one of the great scientific revolutions of the 20th century. It fundamentally changed our understanding of the universe, its history, and our place within it. From a static, eternal cosmos, we now envision a dynamic, evolving universe with a definite beginning and a vast, possibly infinite future.

As we continue to refine and extend the Big Bang model, we push the boundaries of our knowledge ever closer to the ultimate origins of the cosmos. Yet, as with all scientific theories, the Big Bang model is provisional, subject to revision or even replacement as new evidence emerges and our understanding deepens.

In the next chapter, we will turn our attention to a very different conception of cosmic origins: the Hindu concept of Brahman. As we explore this philosophical and spiritual perspective, we will begin to see intriguing parallels and contrasts with the scientific worldview embodied in the Big Bang theory.

19

3. Brahman – The Ultimate Reality in Hindu Philosophy

1. Introduction: The Quest for Ultimate Reality

While modern science has made remarkable strides in understanding the physical universe, as we explored in the previous chapter, humanity's quest to comprehend the nature of reality extends far beyond empirical observation. For millennia, philosophers and spiritual seekers have grappled with questions of existence, consciousness, and the fundamental nature of reality. In the rich tapestry of human thought, few concepts are as profound and far-reaching as the Hindu notion of Brahman.

Brahman, often described as the ultimate reality or absolute consciousness in Hindu philosophy, represents a radically different approach to understanding the cosmos and our place within it. Unlike the Big Bang theory, which seeks to explain the physical origins and evolution of the universe, the concept of Brahman delves into the very nature of existence itself, challenging our conventional notions of reality, consciousness, and the self.

In this chapter, we will explore the concept of Brahman in depth, tracing its development through Hindu thought, examining its key attributes, and considering its implications for our understanding of the cosmos and our own nature. As we delve into this profound philosophical concept, we will begin to see both striking contrasts and intriguing parallels with the scientific worldview we explored in the previous chapter.

2. Historical Development of the Concept of Brahman

The concept of Brahman has a long and rich history in Indian philosophy, evolving and deepening over thousands of years. To understand Brahman, we must trace its development through key texts and schools of thought in the Hindu tradition.

2.1 Vedic Origins

According to historians, the roots of the concept of Brahman can be traced back to the Vedas, the oldest sacred texts of Hinduism, dating back to around 1500-1200 BCE. In the early Vedic period, the term "brahman" referred to the sacred power inherent in rituals and mantras. It was seen as the underlying force that gave efficacy to religious practices.

However it may be clarified that, traditionally Vedas are considered not having any human author hence do not have a beginning. Thus fixing a date to origin of Vedas may not be a good idea.

2.2 The Upanishads: Brahman as Ultimate Reality

The concept of Brahman as the ultimate reality or absolute consciousness emerged in the Upanishads, philosophical texts composed between 800-200 BCE. However,the origin of the Upanishads is a matter of debate, but the most widely accepted theory is that they began as oral traditions of ascetics.

The Upanishads are a part of the Vedas, a larger group of texts. They are considered to be the beginning of a reasoned inquiry into philosophical questions, such as:

- The nature of being

- The nature of the self

- The foundation of life

- What happens to the self at death

- The good life

- Ways of interacting with others

The Upanishads are considered to be the fountainhead of India's philosophical tradition. They contain some of the oldest discussions of philosophical terms, such as:

- ātman (the self)

- brahman (ultimate reality)

- karma

- yoga

- saṃsāra (worldly existence)

- mokṣa (enlightenment)

- puruṣa (person)

- prakṛti (nature)

In the Upanishads, Brahman is described as the eternal, unchanging reality that is the source and ground of all existence. It is often characterized through negation (neti neti - "not this, not this"), as it is considered beyond all attributes and descriptions.

Key Upanishadic statements about Brahman include:

- **"Tat Tvam Asi"** (That Thou Art) - indicating the identity of the individual self (Atman) with Brahman.

- **"Aham Brahmasmi"** (I am Brahman) - asserting the ultimate identity of the self with the absolute.

- **"Prajnanam Brahma"** (Consciousness is Brahman) - equating pure consciousness with the ultimate reality.

2.3 Classical Hindu Philosophy: Six Darshanas

As Hindu philosophy developed, six main schools of thought (darshanas) emerged, each offering its own interpretation of the Vedas and the nature of reality. The concept of Brahman plays a central role in several of these schools:

1. **Vedanta**: Perhaps the most influential school in terms of the concept of Brahman. Vedanta, particularly in its Advaita (non-dualist) form, emphasizes the absolute oneness of Brahman.

2. **Samkhya**: While not directly addressing Brahman, Samkhya's dualistic philosophy of Purusha (consciousness) and Prakriti (matter) influenced later interpretations of Brahman.

3. **Yoga**: Closely related to Samkhya, the Yoga school focuses on the practical methods of realizing the ultimate reality through meditation and other practices.

4. **Nyaya**: Nyāya is a school of Hindu philosophy that focuses on epistemology, metaphysics, and logic.

5. **Vaisheshika**: Vaisheshika, is significant for its naturalism, a feature that is not characteristic of most Indian thought. The Sanskrit philosopher Kanada Kashyapa (probably in 2nd–3rd century CE) expounded its theories and is credited with founding the school. After a period of independence, the Vaisheshika school fused entirely with the Nyaya school, a process that was completed in the 11th century. Thereafter the combined school was referred to as Nyaya-Vaisheshika.

6. **Mimamsa**: Initially focused on Vedic ritual, later Mimamsa thinkers developed sophisticated arguments about the nature of ultimate reality.

2.4 Vedanta and Its Schools

The Vedanta school, based primarily on the interpretation of the Upanishads, became the most influential in developing the concept of Brahman. Several sub-schools of Vedanta emerged, each with its own understanding of the nature of Brahman and its relationship to the world and individual souls:

1. **Advaita Vedanta**: Developed by Adi Shankara (8th century CE), this non-dualist school posits that Brahman alone is real, and the apparent multiplicity of the world is ultimately an illusion (maya).

2. **Vishishtadvaita**: Propounded by Ramanuja (11th century CE), this qualified non-dualist school sees the world and individual souls as real but entirely dependent on and part of Brahman.

3. **Dvaita**: Formulated by Madhva (13th century CE), this dualist school maintains a fundamental distinction between Brahman (identified with Vishnu), individual souls, and the material world.

4. **Bhedabheda**: This school, with various sub-branches, proposes a view of simultaneous difference and non-difference between Brahman and the world.

These different interpretations of Brahman have led to rich philosophical debates within the Hindu tradition, contributing to the depth and complexity of the concept.

3. Key Attributes of Brahman

While different schools of Hindu philosophy have varying interpretations of Brahman, there are several key attributes that are generally associated with this concept:

3.1 Sat-Chit-Ananda

Brahman is often described as Sat-Chit-Ananda, which translates to **Existence-Consciousness-Bliss:**

- **Sat**: Eternal, unchanging existence

- **Chit**: Pure consciousness or awareness

- **Ananda**: Absolute bliss or joy

This triad is seen as the essential nature of Brahman, beyond all attributes and forms.

3.2 Nirguna and Saguna Brahman

Many schools of thought distinguish between two aspects of Brahman:

- **Nirguna Brahman**: Brahman without attributes, the absolute reality beyond all description and conceptualization.

- **Saguna Brahman**: Brahman with attributes, often personified as a deity or seen as the manifest universe.

3.3 Infinity and Eternity

Brahman is considered to be infinite in both space and time. It is beyond the limitations of space-time, encompassing all that exists and transcending all boundaries.

3.4 Non-Duality

In Advaita Vedanta, Brahman is seen as the only reality. The apparent multiplicity of the world is considered to be an illusion (maya) arising from ignorance (avidya) of the true nature of reality. It may be reiterated, there is One Supreme Reality - Brahman which due to Its intrinsic power (maya) appears like the world and the knowers of the world (Jivas). Human knowers due to their ignorance of Brahman as their own self considers the duality as real.

3.5 Self-Luminous

Brahman is often described as self-luminous or self-revealing. It is the light by which all else is known, the consciousness that illuminates all experience.

3.6 The Source and Substratum of All

Brahman is considered to be both the efficient and material cause of the universe. It is the source from which all manifestation arises and the underlying reality that supports all existence.

4. Brahman and the Nature of Reality

The concept of Brahman has profound implications for how we understand the nature of reality. Let's explore some of these implications:

4.1 The Illusion of Multiplicity

In Advaita Vedanta, the apparent diversity and multiplicity of the world are considered to be an illusion (maya) superimposed on the underlying unity of Brahman. This doesn't mean that the world is unreal in an absolute sense, but rather that its apparent separate existence is not ultimately real.

4.2 Consciousness as Fundamental

Unlike materialist philosophies that see consciousness as an emergent property of matter, many schools of Hindu thought posit consciousness (in the form of Brahman) as the fundamental reality. The material world is seen as a manifestation or expression of this consciousness.

4.3 The Nature of the Self

The concept of Brahman is closely tied to the understanding of the self (Atman) in Hindu philosophy. The Upanishads declare the ultimate identity of Atman and Brahman, suggesting that our true nature is not the limited individual ego, but the infinite, eternal consciousness that is Brahman.

4.4 Beyond Space and Time

Brahman is considered to transcend the limitations of space and time. From this perspective, the universe as we perceive it, with its spatial and temporal dimensions, is a limited expression of a reality that is beyond such categories.

Time and Space are not objects of perception. None of our five senses , eyes, ears etc. can sense time and space. They are inferred when objects are observed. The brain processes time and space together, and the brain's mechanisms for measuring them are inferential. The brain doesn't have dedicated sensors for time and space, so it constructs space from parts and relationships.

With respect to objects we get the knowledge like this table is here, that chair is there. 'Here' and 'there' are not any characteristics of the table or chair. They are something different from them but are felt when observing them. These are the inferences of the presence of space.

Similarly we have the inference 'before' 'now' 'afterwards' with respect to objects. Like this Apple is ripe now, it was raw before and it will get spoilt afterwards. The characteristics of now before afterwards are not present in the Apple but they are inferred while observing the Apple.

These are space and time which can be measured.

But there are space and time which cannot be measured. That is the nature of Brahman. Brahman is referred to as Chidākāsha space of consciousness. And of the nature immutable time Akshaya Kala.

With respect to us we always understand wherever our body is as I am here and I am now. Space and time are experienced as here and now with respect to the self.

In deep sleep no objects are experienced hence space and time are also not experienced.

Vedanta says time and space are product of Maya. Maya is the power of Brahman hence not different from Brahman. Hence space and time are products of Brahman. Just as rest of the world is a product of Brahman. Rather the world is an expression of Brahman. The world includes the observers as well as the observed.

Future doesn't exist now. It is the result of our karma. Past exists in the present as samskars. Brahman exists in all time, in all space, in all objects. Brahman exists here and now.

4.5 The Problem of Causation

If all effects are the result of previous causes, then the cause of a given effect must itself be the effect of a previous cause, which itself is the effect of a previous cause, and so on, forming an infinite logical chain of events that can have no beginning.

The relationship between Brahman and the manifest world raises complex philosophical questions about causation. How does the unchanging, eternal Brahman give rise to the changing, temporal world? Various schools of thought have proposed different solutions to this problem, from the concept of vivarta (apparent transformation) in Advaita to the idea of real transformation (parinama) in some other schools.

5. Paths to Realizing Brahman

Hindu philosophy doesn't merely posit Brahman as a theoretical concept but suggests various paths for directly realizing this ultimate reality. These paths, often intertwining, include:

5.1 Jnana Yoga: The Path of Knowledge

This path emphasizes intellectual discrimination and self-inquiry to pierce through the veil of ignorance and realize one's true nature as Brahman.

5.2 Bhakti Yoga: The Path of Devotion

Through devotion to a personal form of the divine (Saguna Brahman), the practitioner ultimately realizes the impersonal absolute (Nirguna Brahman).

5.3 Karma Yoga: The Path of Selfless Action

By performing actions without attachment to their fruits, one can transcend the limited ego and realize the underlying unity of all.

5.4 Raja Yoga: The Path of Meditation

Through various meditative practices, one can directly experience the pure consciousness that is Brahman.

6. Brahman and the Cosmos

The concept of Brahman offers a unique perspective on the nature and origin of the cosmos:

6.1 Emanation vs. Creation

Unlike the Abrahamic concept of creation ex nihilo (out of nothing), many Hindu cosmologies see the universe as an emanation or manifestation of Brahman. The universe is not separate from Brahman but is a expression of it.

6.2 Cycles of Manifestation

Many Hindu traditions speak of cosmic cycles of manifestation and dissolution. The universe is seen as periodically emerging from and dissolving back into Brahman.

6.3 The Nature of Time

From the perspective of Brahman, which is eternal and beyond change, time itself is often seen as part of the cosmic illusion. The linear progression of time is considered a feature of our limited perception rather than an absolute reality.

7. Philosophical Challenges and Debates

The concept of Brahman, while profound, has faced various philosophical challenges and sparked numerous debates:

7.1 The Problem of Evil

If Brahman is all-encompassing and the source of everything, how do we account for the existence of evil and suffering in the world?

7.2 The Relation Between Brahman and the World

Different schools of thought have debated whether the world is real and different from Brahman, real but non-different, or ultimately unreal.

7.3 The Nature of Liberation

Is liberation (moksha) the realization of an already existing identity with Brahman, or is it a state to be achieved?

7.4 The Role of Reason

Can Brahman be known through reason and intellectual inquiry, or is it beyond the reach of the conceptual mind?

8. Brahman and Modern Thought

The concept of Brahman continues to be relevant in modern philosophical and scientific discussions:

8.1 Parallels with Quantum Physics

Some thinkers have drawn parallels between the non-dual nature of Brahman and the interconnectedness suggested by quantum physics. Reader may refer to Appendice F and G for more details.

8.2 Consciousness Studies

As the study of consciousness becomes more prominent in science and philosophy, the idea of consciousness as fundamental (as in the concept of Brahman) is receiving renewed attention.

8.3 Holistic Worldviews

The all-encompassing nature of Brahman resonates with holistic and systems-based approaches in various fields, from ecology to psychology.

9. Conclusion: Brahman as a Philosophical Revolution

The concept of Brahman represents a profound philosophical revolution, challenging our ordinary notions of reality, self, and the cosmos. It offers a vision of ultimate unity underlying all apparent diversity, of consciousness as the fundamental reality, and of our own deepest nature as identical with the absolute.

As we move forward in our exploration, we will begin to see how this ancient philosophical insight can enter into dialogue with modern scientific cosmology. While the Big Bang theory and the concept of Brahman arise from very different modes of inquiry, both seek to understand the fundamental nature of our cosmos. In the chapters that follow, we will explore the intriguing parallels and contrasts between these two perspectives, seeking a more comprehensive understanding of our cosmic origins and the nature of reality itself.

4. Time and Eternity

1. Introduction: The Enigma of Time

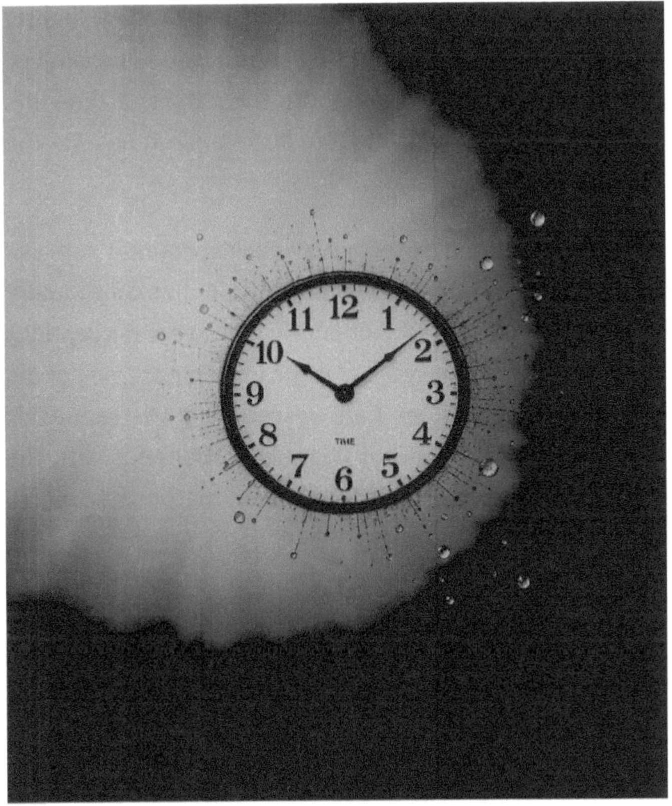

Time is perhaps the most pervasive yet elusive aspect of our existence. It shapes our experiences, governs our lives, and fundamentally structures our understanding of the universe. Yet, when we try to grasp its true nature, time seems to slip through our fingers like sand. Is time a fundamental feature of the universe, or merely a construct of human perception? Does it flow inexorably forward, or is the passage of time an illusion? Is there a beginning or end to time, or is it eternal?

In our exploration of cosmic origins, the nature of time takes center stage. The Big Bang theory, as we discussed in Chapter 2, presents a vision of the universe with a definite beginning in time. In contrast, the Hindu concept of Brahman, explored in Chapter 3, often suggests a reality beyond time, eternal and unchanging. In this chapter, we will delve deep into the nature of time and eternity, examining perspectives from both modern physics and Hindu philosophy.

As we navigate this complex terrain, we will encounter some of the most profound and perplexing questions in human thought: What is the fundamental nature of time? How do our perceptions of time relate to its physical reality? Is the universe eternal, or did time itself have a beginning? Can we reconcile the seemingly linear progression of time in our everyday experience with the cyclical time concepts in Hindu thought or the block universe suggested by relativity theory?

By exploring these questions from both scientific and philosophical perspectives, we aim to develop a richer, more nuanced understanding of time and its role in our conceptions of cosmic origins.

2. Time in Modern Physics

Our scientific understanding of time has undergone radical transformations over the past century. Let's explore some key developments in the physics of time:

2.1 Newtonian Absolute Time

For centuries, the prevailing scientific view of time aligned closely with our intuitive understanding. Isaac Newton conceived of time as absolute, flowing uniformly without relation to anything external. In this view, time was a universal backdrop against which all events unfolded.

2.2 Einstein's Relativity: The Demise of Absolute Time

Albert Einstein's theories of special and general relativity revolutionized our understanding of time:

2.2.1 Special Relativity

Special relativity revealed that time is not absolute but relative to the observer's frame of reference. Key implications include:

- Time Dilation: Time passes more slowly for objects moving at high velocities relative to the observer.

- Simultaneity is Relative: Events that appear simultaneous to one observer may not be simultaneous to another.

- Spacetime: Time and space are intertwined in a four-dimensional continuum called spacetime.

2.2.2 General Relativity

General relativity further revealed that time is affected by gravity:

- Gravitational Time Dilation: Time passes more slowly in stronger gravitational fields.

- Curvature of Spacetime: Massive objects curve spacetime, affecting the passage of time around them.

- Black Holes: At the event horizon of a black hole, time (from an outside observer's perspective) appears to stop altogether.

2.3 The Arrow of Time

While the fundamental laws of physics are largely time-symmetric, our universe displays a clear "arrow of time":

- Thermodynamic Arrow: The tendency for entropy to increase in closed systems.

- Cosmological Arrow: The expansion of the universe.

- Psychological Arrow: Our perception of time as flowing from past to future.

The origin and nature of time's apparent directionality remain subjects of intense scientific and philosophical debate.

2.4 Quantum Time

Quantum mechanics introduces further complications to our understanding of time:

- Quantum Superposition: Particles can exist in multiple states simultaneously until observed.

- The Measurement Problem: The act of measurement appears to "collapse" the wave function, raising questions about the nature of time in quantum systems. Reader may refer to observer effect in Appendix G.

- Wheeler-DeWitt Equation: Some formulations of quantum gravity suggest that time may be an emergent property rather than a fundamental feature of reality.

2.5 The Beginning of Time?

The Big Bang theory suggests that our universe, including time itself, had a beginning. However, this raises profound questions:

- What caused the Big Bang?

- Can we meaningfully talk about "before" the Big Bang?

- Are there other models that avoid a beginning of time?

Some theories, like eternal inflation or cyclic universe models, propose alternatives to a single beginning of time.

3. Time in Hindu Philosophy

Hindu thought offers a rich and multifaceted understanding of time, often strikingly different from Western conceptions:

3.1 Cosmic Cycles: Yugas and Kalpas

Hindu cosmology often describes time in vast cycles:

- Yugas: Four ages (Satya, Treta, Dvapara, and Kali) that repeat in a cycle.

- Maha Yuga: A complete cycle of four Yugas.

- Kalpa: A day in the life of Brahma, equivalent to 4.32 billion years.

- The breath of Brahma: Cycles of creation and dissolution of the universe.

These cycles suggest a view of time that is neither linear nor finite, but cyclical and eternal.

3.2 Maya and the Illusion of Time

In Advaita Vedanta, time is often considered part of maya, the illusory nature of the phenomenal world:

- Time as a construct of the mind, not an absolute reality.

- The "eternal now" as the true nature of reality.

- Transcendence of time as a key aspect of spiritual realization.

3.3 Kaala: Time as a Principle of Change

Some schools of Hindu thought personify time as Kaala, a principle of change and transformation:

- Kaala as a force that brings about the manifestation and dissolution of the universe.

- The dual nature of time as both creative and destructive.

3.4 Time and Karma

The concept of karma introduces a moral dimension to time:

- Actions in one lifetime having consequences in future lives.

- The role of time in the unfolding of karmic effects.

3.5 Timelessness of Brahman

As discussed in the previous chapter, Brahman is often described as beyond time:

- Eternal, unchanging nature of ultimate reality.

- Time as a feature of the manifest world, not of Brahman itself.

4. Comparative Analysis: Physics and Hindu Concepts of Time

Despite their different origins and methodologies, modern physics and Hindu philosophy offer some intriguing parallels in their conceptions of time:

4.1 Relativity of Time

- Physics: Time is relative to the observer's frame of reference.

- Hindu Thought: Time is often seen as subjective, with different realms of existence experiencing time differently.

4.2 Timelessness

- Physics: Some theories of quantum gravity suggest a timeless underlying reality.

- Hindu Thought: Brahman is described as beyond time.

4.3 Cyclic Time

- Physics: Some cosmological models propose cyclic universes.

- Hindu Thought: Cosmic cycles of creation and dissolution.

4.4 Emergent Time

- Physics: Time may be an emergent property rather than fundamental.

- Hindu Thought: Time as part of maya, not ultimate reality.

4.5 Multiple Timescales

- Physics: Different processes in the universe operate on vastly different timescales.

- Hindu Thought: Yugas and Kalpas describe cosmic time on various scales.

5. Philosophical Implications

The nature of time raises profound philosophical questions:

5.1 The Reality of Time

Is time a fundamental feature of the universe, or is it in some sense illusory?

Whether time is a fundamental feature of the universe or an illusion is a topic of debate among physicists and scientists:

- Some say time is an illusion

Some physicists believe time is not fundamental and should be removed from physics equations. They say that time is a function of our relationship with the physical world and changes based on our psychological and physical states. Others say that time is a human construct, a result of neurons firing and memories formed.

- Others say time is fundamental

Some say that time is an object with physical size and that the universe expands in time. They say that time is an ever-moving fabric through which we experience things coming together and apart.

- It's not clear which is fundamental

It's not clear if time is a fundamental or emergent property of the universe. We can't currently test which one is fundamental and which is emergent.

- Space and time are illusions

Some physicists, like Nathan Seiberg, believe that space and time are illusions and are products of a more fundamental projector reality.

5.2 Free Will and Determinism

How does our understanding of time affect notions of free will and determinism?

The way time is understood can affect the debate over free will and determinism in a number of ways, including:

- Defining determinism

The definition of determinism can be made simpler and more relevant by considering the temporal structure of possibilities.

- Understanding the future

Determinism is the idea that the future is determined by the present and past, and that later states are a result of earlier states. However, it's possible for determinism to be true even if the future is unpredictable.

- The role of indeterminism

Incorporating a model of indeterminism into scientific discussions about determinism could help to dispel the idea that determinism is the only scientific picture of the world.

- The scope of determinism

Some philosophers believe that the universe is a single determinate system, while others believe that there are more limited determinate systems.

- Compatibilism and incompatibilism

These are opposing sides of the debate over whether determinism and free will can coexist.

Free will is the idea that humans have the ability to make decisions and act independently of any previous event or state of the universe. It's a controversial topic that's been debated in philosophy, science, and theology.

5.3 The Nature of Change

If reality is timeless, how do we account for the apparent change and becoming in our experience?

If reality is timeless, the apparent change and becoming in our experience can be explained by the idea that time is an emergent property of the world's basic ingredients.

Here are some other ideas about time and reality:

- Timeless universe

This theory rejects the reality of time, arguing that it's a human illusion. In a timeless universe, the cosmos is eternal and all physical processes operate within a timeless framework.

- Eternalist block-universe

This view suggests that the apparent flow of time from past to future is an illusion. It's based on the idea that the universe comes into being moment by moment.

- Time is not real

Some scientists believe that time is not real and should be removed from the equations of physics.

- Time is a dimension of reality

Others say that time is a dimension of reality and that the future does not exist until we create it.

- Timelessness is the transcendence of time

Timelessness is the state of transcending time, where there is no awareness of time at all.

5.4 Eternity and Immortality

What do different conceptions of time imply about the possibility of eternal existence or immortality?

Different conceptions of time can have different implications for the possibility of eternal existence or immortality, including:

- Eternalism

This philosophical approach views all existence in time as equally real, which is different from presentism or the growing block universe theory of time.

- Karl Jaspers

Jaspers believed that humans are both mortal and immortal, and that immortality can only be achieved in the present, while living life in the face of death.

- Poincaré recurrence theorem

This theorem states that certain dynamical systems will return to a state similar to their original state infinitely often. Some argue that this could prove reincarnation.

- Secular futurists

Some envision technologies that could suspend death indefinitely, such as mind uploading and Strategies for Engineered Negligible Senescence.

- Philosophical and religious thought

Immortality can take many forms, including bodily immortality, soul immortality, or symbolic immortality.

5.5 The Meaning of Now

What is the ontological status of the present moment?

The ontological status of the present moment is debated in philosophy, with different views including presentism, A-theory, and B- and C-theories:

- Presentism: The view that only the present exists, and that the past and future do not. It's often described as the idea that "to be present is to be real".
- A-theory: The view that there is an objective present moment that changes over time.
- B- and C-theories: The view that the present is not changing, and that times are ordered but directionless.

Other views on the ontological status of time include:

- Eternalism: The view that past, present, and future entities exist.
- No-futurism: The view that only past and present entities exist.

6. Time and Consciousness

The relationship between time and consciousness is a crucial area where scientific and spiritual perspectives intersect:

6.1 The Experience of Time

How does our subjective experience of time relate to its physical nature?

Our subjective experience of time is a mental representation of physical time, and it's actively generated by the brain and body:

- Mental representation

Subjective time is the conscious or unconscious mental representation of physical time.

- Brain and body

The brain's predictive processes and the body's sensorimotor activities actively generate our experience of time.

- Factors that influence time perception

Time perception is subjective and influenced by many factors, including attention, memory, emotion, and physical status. For example, time may seem to stand still when you're stuck in traffic, but fly by when you're having fun.

- Different neural systems

Different neural systems are involved in temporal processes depending on the duration of the processed interval.

- Subjective experience across species

There's no guarantee that there are characteristic differences in the subjective experience of time across species.

6.2 Altered States of Consciousness

How do altered states (e.g., meditation, psychedelics) affect our perception of time?

Altered states of consciousness, such as those that occur during meditation or drug use, can affect our perception of time in a number of ways:

- Meditation: During peak meditation, people may experience a dissolution of their sense of self and time, leading to feelings of "timelessness" and "selflessness". Meditation can also cause the frontal cortex, which processes sensory information and helps orient us in time and space, to go offline.
- Drugs: Some psychoactive drugs can cause hallucinogen-persisting perception disorders, also known as "flashbacks", which can occur days or weeks after the effects of the drug have worn off.
- Relaxation: Increased relaxation can be associated with a slower perception of time.
- Mindfulness: Mindfulness is associated with a more mindful, slower passage of time.

6.3 Memory and Anticipation

How do our capacities for remembering the past and anticipating the future shape our experience of time?

The ability to remember the past and anticipate the future shapes our experience of time in several ways, including:

- Understanding the present: Memories help us understand the present by drawing on past events to frame our behavior and understanding.
- Predicting the future: Memories help us predict what might happen in the future.

- Planning: Memories help us plan for the future by allowing us to mentally try out different strategies and work through potential outcomes.
- Sense of self: Memory is critical for our sense of who we are.
- Mental time travel: The ability to mentally re-experience a past event is known as mental time travel.

The ability to remember personally experienced events from the past is called episodic memory. Remembering past events is thought to have evolved to help with future decision making.

6.4 The "Specious Present"

The notion that our experience of the present is actually an integration of a short duration of time, not an infinitesimal instant.

William James famously wrote that the "specious present" is "the short duration of which we are immediately and incessantly sensible" (1886, 397). He drew on the 1870s–1880s work of Shadworth Hodgson and Robert Kelly, who have been described as the "independent inventors"[1] of the specious present theory.

6.5 Timelessness in Mystical Experiences

Reports of timelessness in mystical experiences across cultures and their potential implications.

Timelessness is a characteristic of mystical experiences:

- Definition

Mystical experiences are transient, extraordinary experiences that can include feelings of timelessness and spacelessness.

- Characteristics

Other characteristics of mystical experiences include:

- A sense of unity with all things
- A sense of being in a harmonious relationship with the divine
- Euphoria
- Loss of ego functioning
- Alterations in time and space perception
- Nature

Mystical experiences are considered to be human experiences that are potentially available to everyone. They are not dreaming or hallucinations, but they do differ from the emotional, intellectual, and volitional life of ordinary people.

- Hypotheses

There are two competing hypotheses regarding mystical experiences:

- Temporal Involvement Hypothesis: Mystical experiences are induced by activity in regions of the brain associated with emotion, abstract semantics, and imagery.
- Executive Inhibition Hypothesis: Mystical experiences are underpinned by executive down-regulation in response to authoritative suggestions.

7. Eternity: Beyond Time

The concept of eternity presents unique challenges to both scientific and philosophical thought:

7.1 Eternity as Infinite Duration

Eternity is the idea of an infinite amount of time that never ends, or the quality of being everlasting. In classical philosophy, eternity is defined as something that exists outside of time or is timeless.

Here are some related ideas about eternity:

- Time and eternity

Time is based on change, while eternity is unchanging.

- Measuring eternity

Eternity is something that cannot be conceived of or measured, so the word "forever" is used to describe it.

- Using the word "eternity"

The word "eternity" can also be used to describe a very long period of time. For example, people who promise to love each other for eternity are not planning to ever split up.

7.2 Eternity as Timelessness

The notion of eternity as a state or reality beyond or outside of time altogether.

Eternity is the idea of a state of existence outside of time or a time without end. It is a concept that is important in many religions and lives, and is often used to describe the existence of gods or an afterlife.

Here are some related ideas about eternity:

- The Theory of Eternity of Life

This philosophical concept suggests that life is eternal and infinite, and that the soul continues to exist after death.

- Plato's conception of Heaven

Plato believed that true reality is present in an unchanging Heaven, and that the Forms or Ideas can be glimpsed by the trained intellect.

- Boethius's concept of eternity

Boethius believed that eternity is the whole, simultaneous, and perfect possession of boundless life.

- The history of the word eternity

The word eternity comes from the Latin term aeternitas, and the Greek word αἰών.

- Parmenides

Some consider Parmenides to be the first philosopher to articulate a notion of eternity.

7.3 Eternal Return

Nietzsche's concept of eternal return and its relation to both scientific and Hindu cyclical time concepts.

Friedrich Nietzsche's concept of eternal return is related to scientific and Hindu cyclical time concepts in a few ways:

- Eternal return

Nietzsche's idea of eternal return is the idea that life will repeat infinitely, with every detail the same in each life. He believed that all that has happened will happen again, and that we have the ability to choose what will recur.

- Scientific cyclical time

An example of scientific cyclical time is the repeating seasons. Early agricultural societies were centered around the seasons, the sun's rising and falling, and the changing star formations.

- Hindu cyclical time

In Hindu cosmology, time is eternal and repeats in four types of cycles. The smallest cycle is a maha-yuga, which contains four yugas. A manvantara contains 71 maha-yugas. The term kāla chakra is often used in Hindu conversation to refer to the cycle of time

7.4 Eternity and Cosmology

How do different cosmological models (e.g., Big Bang, steady state, cyclic universes) relate to the concept of eternity?

The Big Bang, steady state, and cyclic universe cosmological models differ in how they relate to the concept of eternity, as follows:

- Big Bang

The Big Bang theory states that the universe had a beginning and a finite age. It describes how the universe expanded from a state of infinite density, and how matter and radiation populated the universe with particles, atoms, stars, and galaxies.

- Steady state

The steady state theory states that the universe is infinite, with no beginning or end. It proposes that the universe is always expanding, but that new matter is continuously created to maintain a constant density. This theory is based on the idea that the universe is homogeneous and isotropic in space and time, and that it looks the same from any place or time. However, most cosmologists, astrophysicists, and astronomers reject the steady state theory in favor of the Big Bang theory.

- Cyclic universe

The cyclic universe theory proposes that the universe expands and then contracts, leading to the formation of a new universe. Some believers in this theory have published scientific papers claiming to have found evidence of concentric circles and other features that would support the theory. However, most astronomers believe that these features are not real, and that the general view is that there is no evidence to support the cyclical universe theory.

7.5 Eternity and Ultimate Reality

The relationship between eternity and concepts of ultimate reality in various philosophical and spiritual traditions.

The relationship between eternity and ultimate reality varies across philosophical and spiritual traditions, with some traditions conceiving of ultimate reality as an eternal truth, while others conceive of it as an absolute state of being:

- Hinduism

The ultimate reality in Hinduism is Brahman, which is an eternal truth that is the cause of all change. Brahman is both immanent (manifested in the present) and transcendent (beyond reach).

- Taoism

The ultimate reality in Taoism is the Tao, which is an impersonal eternal truth.

- Buddhism

The ultimate reality in Buddhism varies by school, but can include Shunyata and Nirvana.

- Islam

The ultimate reality in Islam is Allah, which is similar to the concept of God in the Bible.

- Judeo-Christian

The ultimate reality in Judeo-Christian thought is a personal God who created the universe.

- Indian philosophy

In Indian philosophy, the soul, or ātman, is directly connected to the ultimate reality. Indian philosophy also holds that the soul is central to understanding epistemology.

- Advaita Vedanta

In Advaita Vedanta, the ultimate reality is an absolute state of being that cannot be described by attributes like omniscience or omnipotence.

8. Time, Eternity, and Human Existence

Our understanding of time and eternity profoundly affects how we view our own existence and its meaning:

8.1 The Human Lifespan

How do we reconcile our finite lifespan with concepts of eternal time or timeless reality?

8.2 The Search for Meaning

How do different conceptions of time and eternity affect our search for meaning and purpose?

8.3 Ethics and Time

How does our understanding of time influence our ethical frameworks and decision-making?

8.4 Death and Immortality

How do various perspectives on time relate to our understanding of death and notions of immortality?

8.5 Living in Time

Practical and existential implications of different time concepts for how we live our lives.

9. Conclusion: Towards a Holistic Understanding of Time and Eternity

As we've explored in this chapter, the concepts of time and eternity are central to both scientific cosmology and spiritual philosophies like Hinduism. While modern physics and Hindu thought approach these concepts from very different perspectives, both challenge our everyday notions of time and point towards a deeper, more complex reality.

The linear, absolute time of our everyday experience gives way, under scrutiny, to a more fluid, relative, and perhaps even illusory nature. Both relativistic physics and Advaitic philosophy suggest that our common-sense notions of time may be more a feature of our limited perception than a fundamental aspect of reality.

Yet, the apparent flow of time and the distinction between past, present, and future remain central to our lived experience. The challenge, then, is to integrate these diverse perspectives into a more comprehensive understanding - one that can account for both the time-bound nature of our everyday existence and the possibility of a timeless, eternal reality.

As we continue our exploration of cosmic origins, bridging the Big Bang and Brahman, this multifaceted understanding of time will be crucial. It allows us to appreciate both the temporal unfolding of the universe described by modern cosmology and the timeless, eternal nature of ultimate reality posited in Hindu philosophy.

In the next chapter, we will delve into nature of reality that would take us, later, to the nature of consciousness and its role in the cosmos.

5. The Nature of Reality

1. Introduction: Peering Behind the Veil

What is the true nature of reality? This question has captivated human minds for millennia, driving both scientific inquiry and philosophical contemplation. As we delve deeper into our exploration of cosmic origins, we find ourselves confronting this fundamental question head-on. The nature of reality underlies our understanding of the universe's beginning, its ongoing existence, and our place within it.

In this chapter, we will examine two profound yet seemingly disparate perspectives on the nature of reality: the view emerging from modern quantum mechanics and the concept of Maya in Hindu philosophy, particularly as articulated in Advaita Vedanta. Despite their vastly different origins and methodologies, these perspectives offer intriguing parallels in their challenge to our conventional understanding of reality.

Quantum mechanics, one of the most successful scientific theories ever developed, presents a picture of reality that is probabilistic, interconnected, and often counterintuitive. It suggests a world where particles can exist in multiple states simultaneously, where measurement affects reality, and where the very act of observation plays a crucial role in determining what we perceive as real.

On the other hand, the Hindu concept of Maya, often translated as "illusion" or "appearance," suggests that the reality we perceive with our senses is not the ultimate truth. Instead, it posits that there is a deeper, unchanging reality (Brahman) underlying the ever-changing world of appearances.

As we explore these perspectives, we will grapple with profound questions: Is reality fundamentally deterministic or probabilistic? Is the world we perceive through our senses the ultimate reality, or is there a deeper truth beyond appearances? How do consciousness and observation relate to the nature of reality?

By examining these questions through both scientific and philosophical lenses, we aim to develop a richer, more nuanced understanding of the nature of reality and its implications for our conception of cosmic origins.

2. Quantum Mechanics: Reality at its Most Fundamental

Quantum mechanics, developed in the early 20th century, revolutionized our understanding of the physical world at its most fundamental level. Let's explore key aspects of quantum theory and their implications for the nature of reality:

2.1 The Wave-Particle Duality

One of the most striking features of quantum mechanics is the wave-particle duality of matter and energy:

- Light, traditionally thought of as a wave, can behave like particles (photons).

- Particles like electrons can exhibit wave-like properties.

- The famous double-slit experiment demonstrates this dual nature.

This duality challenges our classical notions of distinct particles and waves, suggesting a more fluid and interconnected reality at the quantum level.

Wave-particle duality, possession by physical entities (such as light and electrons) of both wavelike and particle-like characteristics. On the basis of experimental evidence, German physicist Albert Einstein first showed (1905) that light, which had been considered a form of electromagnetic waves, must also be thought of as particle-like, localized in packets of discrete energy.

2.2 The Uncertainty Principle

Heisenberg's Uncertainty Principle states that certain pairs of physical properties, such as position and momentum, cannot be simultaneously measured with arbitrary precision:

- This is not merely a limitation of measurement but a fundamental feature of reality.

- It suggests that at the quantum level, reality is inherently probabilistic rather than deterministic.

In everyday life, calculating the speed and position of a moving object is relatively straightforward. We can measure a car traveling at 60 miles per hour or a tortoise crawling at 0.5 miles per hour and simultaneously pinpoint where the objects are located. But in the quantum world of particles, making these calculations is not possible due to a fundamental mathematical relationship called the uncertainty principle.

Formulated by the German physicist and Nobel laureate Werner Heisenberg in 1927, the uncertainty principle states that we cannot know both the position and speed of a particle, such as a photon or electron, with perfect accuracy; the more we nail down the particle's position, the less we know about its speed and vice versa.

In other words, if we could shrink a tortoise down to the size of an electron, we would only be able to precisely calculate its speed *or* its location, not both at the same time.

Though the Heisenberg uncertainty principle is famously known in quantum physics, a similar uncertainty principle also applies to problems in pure math and classical physics—basically, any object with wave-like properties will be affected by this principle. Quantum objects are special because they all exhibit wave-like properties by the very nature of quantum theory.

To understand the general idea behind the uncertainty principle, think of a ripple in a pond. To measure its speed, we would monitor the passage of multiple peaks and troughs. The more peaks and troughs that pass by, the more accurately we would know the speed of a wave—but the less we would be able to say about its position. The location is spread out among the peaks and troughs. Conversely, if we wanted to know the exact position of one peak of a wave, we would have to monitor just one small section of the wave and would lose information about its speed. In short: the uncertainty principle describes a trade-off between two complementary properties, such as speed and position.[2]

2.3 Quantum Superposition

Quantum superposition is a fundamental principle of quantum mechanics that states that linear combinations of solutions to the Schrödinger equation are also solutions of the Schrödinger equation.

Quantum systems can exist in multiple states simultaneously until measured:

- Schrödinger's famous cat thought experiment illustrates this concept.

---

[2] *Caltech Science Exchange* / *Topics* / *Quantum Science and Technology* / *Uncertainty Principle*

In Schrodinger's imaginary experiment, you place a cat in a box with a tiny bit of radioactive substance. When the radioactive substance decays, it triggers a Geiger counter which causes a poison or explosion to be released that kills the cat. Now, the decay of the radioactive substance is governed by the laws of quantum mechanics. This means that the atom starts in a combined state of "going to decay" and "not going to decay". If we apply the observer-driven idea to this case, there is no conscious observer present (everything is in a sealed box), so the whole system stays as a combination of the two possibilities. The cat ends up both dead and alive at the same time. Because the existence of a cat that is both dead and alive at the same time is absurd and does not happen in the real world, this thought experiment shows that wavefunction collapses are not just driven by conscious observers.[3]

- Superposition challenges our notions of definite states and binary logic.

2.4 Quantum Entanglement

Entanglement occurs when particles interact in ways such that the quantum state of each particle cannot be described independently:

- Einstein referred to this as "spooky action at a distance."

- Entanglement suggests a deep interconnectedness in the fabric of reality.

Quantum entanglement is the phenomenon of a group of particles being generated, interacting, or sharing spatial proximity in such a way that the quantum state of each particle

---

[3] *https://www.wtamu.edu*

of the group cannot be described independently of the state of the others, including when the particles are separated by a large distance. The topic of quantum entanglement is at the heart of the disparity between classicle physics and quantum physics: entanglement is a primary feature of quantum mechanics not present in classical mechanics.[4]

2.5 The Role of the Observer

In quantum mechanics, the act of observation plays a crucial role:

- The measurement problem: how and why quantum measurements have definite outcomes.

- The 'Copenhagen interpretation' suggests that observation "collapses" the wave function. While "Copenhagen" refers to the Danish city, the use as an "interpretation" was apparently coined by Heisenberg during the 1950s to refer to ideas developed in the 1925–1927 period, glossing over his disagreements with Bohr.Thus The Copenhagen interpretation is **a collection of views about the meaning of quantum mechanics**, stemming from the work of Niels Bohr, Werner Heisenberg. Consequently, there is no definitive historical statement of what the interpretation entails.

Over the years, there have been many objections to aspects of Copenhagen-type interpretations, including the discontinuous and stochastic nature of the "observation" or "measurement" process, the difficulty of defining what might count as a measuring device, and the seeming reliance upon classical physics in describing such devices.

---

[4] *https://en.wikipedia.org/wiki/Quantum_entanglement*

- This raises profound questions about the role of consciousness in shaping reality.[5]

2.6 Quantum Field Theory

Quantum field theory (QFT) is a mathematical and conceptual framework that combines quantum mechanics, special relativity, and classical field theory. It's used to describe the behavior of subatomic particles and their interactions with force fields. QFT is a sophisticated and profound theory that's used in many areas of physics.

Quantum field theory, which reconciles quantum mechanics with special relativity, presents reality as composed of fields rather than particles:

- Particles are seen as excitations of these underlying fields.

- This view further blurs the line between matter and energy, substance and relation.

3. Interpretations of Quantum Mechanics

The mathematical formalism of quantum mechanics is well-established, but its interpretation remains a subject of debate. Various interpretations offer different perspectives on the nature of reality:

3.1 Copenhagen Interpretation

- Reality is probabilistic.

- Measurement causes the wave function to collapse.

---

[5] *https://en.wikipedia.org/wiki/Copenhagen_interpretation*

- There is no underlying deterministic reality.

3.2 Many-Worlds Interpretation

- Every possible outcome of a quantum event occurs, each in a different parallel universe.

- Suggests a vast multiverse of parallel realities.

3.3 Pilot Wave Theory (de Broglie-Bohm Theory)

- Proposes an underlying deterministic reality guided by a "pilot wave."

- Maintains non-locality and challenges the need for wave function collapse.

3.4 Quantum Decoherence

- Explains the appearance of wave function collapse through interactions with the environment.

- Bridges the quantum and classical realms.

3.5 QBism (Quantum Bayesianism)

- Interprets the wave function as representing subjective information rather than objective reality.

- Emphasizes the role of the observer in shaping reality.

4. Maya: The Veil of Illusion in Hindu Philosophy

The concept of Maya is central to many schools of Hindu philosophy, particularly Advaita Vedanta. Let's explore its key aspects:

4.1 Maya as Illusion

- Maya is often translated as "illusion" or "appearance."

- It suggests that the world we perceive is not the ultimate reality.

4.2 Maya and Brahman

- Maya is the power of Brahman that veils the true nature of Brahman (ultimate reality).

- It creates the appearance of multiplicity and change in what is ultimately one and unchanging.

4.3 Maya as Creative Power

- Maya is not merely negative or deceptive but also the creative power of the universe.

- It is through Maya that the formless Brahman appears as the formed universe.

4.4 Levels of Reality

Many Hindu philosophers distinguish between different levels of reality:

- Paramarthika Satya: Absolute reality (Brahman)

- Vyavaharika Satya: Empirical or transactional reality (the world of everyday experience)

- Pratibhasika Satya: Illusory reality (like dreams or misconceptions)

4.5 Maya and Ignorance (Avidya)

- Maya is closely related to the concept of Avidya (ignorance) but they are different. While Maya is intrinsic power of Brahman, ignorance is related to the human understanding of the world.

- It is our ignorance of the true nature of reality that keeps us bound to the illusory world.

4.6 Transcending Maya

- The goal of spiritual practice in many Hindu traditions is to see through the veil of Maya.

- This involves realizing one's true nature as identical with Brahman.

Brahman is the One in whom the world is born, is sustained and again merge back. World is not different from Brahman but Brahman is different from the world. Their relationship is like that of Gold and ornaments.

By its own intrinsic power called Maya Brahman appears like the world with seemingly different characteristics then itself. Brahman is Existence, Knowledge, Infinite. World as seen is changing, inert and having finite objects. The unseen Brahman always exists in the world as gold always exists in gold ornaments.

World is both the infinite observers (conscious beings) and the observed insentient/inanimate objects.

The observers (humans) have ignorance about Brahman. They view the observed world as different from themselves. They have misunderstandings about the world and themselves.

Maya is a mysterious inexplicable power of Brahman. Avidya is the ignorance of the observer. Maya and Avidya are different.

5. Comparative Analysis: Quantum Reality and Maya

Despite their vastly different origins and contexts, quantum mechanics and the concept of Maya offer some intriguing parallels:

5.1 Reality Beyond Appearances

- Quantum mechanics suggests a reality fundamentally different from our everyday classical experience.

- Maya posits a ultimate reality (Brahman) beyond the world of appearances.

5.2 The Role of the Observer

- In quantum mechanics, the observer plays a crucial role in determining reality.

- In Advaita Vedanta, the ultimate observer (Atman/Brahman) is the substratum of all experience.

5.3 Interconnectedness

- Quantum entanglement suggests a deep interconnectedness in reality.

- The concept of Maya implies that all apparent diversity is ultimately unified in Brahman.

5.4 Limits of Knowledge

- The uncertainty principle in quantum mechanics sets fundamental limits on what can be known.

- Maya implies limits to our ability to perceive ultimate reality through ordinary means.

5.5 Levels of Reality

- Some interpretations of quantum mechanics suggest different levels of reality (e.g., quantum vs. classical).

- Hindu philosophy often describes different levels of reality (Paramarthika, Vyavaharika, Pratibhasika).

6. Philosophical Implications

The perspectives offered by quantum mechanics and the concept of Maya raise profound philosophical questions:

6.1 The Nature of Consciousness

- How does consciousness relate to the fundamental nature of reality?

The relationship between consciousness and the fundamental nature of reality is a complex topic with multiple theories and ideas.[6]

- Is consciousness a product of material processes, or is it more fundamental?

According to philosophical materialism, mind and consciousness are caused by physical processes, such as the neurochemistry of the human brain and nervous system, without which they cannot exist.[7]

---

[6] *www.sciencedirect.com/science/article/pii/0025556470900465*

[7] *https://en.wikipedia.org/wiki/Materialism*

6.2 Free Will and Determinism

- Does the probabilistic nature of quantum mechanics leave room for free will?

Whether quantum mechanics allows for free will is a topic of debate, with some arguing that it does and others arguing that it does not.[8]

- How does the concept of Maya relate to questions of free will and destiny?

Maya provides us with free will, and this free will can seem to be true in a lower level of existential reality. However, at a higher level of understanding, a wise man will see that all talk of action, consequences, and free will is an illusion.[9]

6.3 The Limits of Knowledge

- Are there fundamental limits to what we can know about reality?

Science can't provide a complete explanation of reality because of fundamental limits to its knowledge. For example, the uncertainty principle and non-linear dynamics suggest that reality is indeterminate. Also, Gödel's theorems show that some mathematical theorems are true but can't be proven.[10]

---

[8] https://arxiv.org/pdf/quant-ph/0208104

[9] https://www.linkedin.com/pulse/hindu-idea-destiny-free-vinod-aravindakshan-tbi7c/

[10] https://en.unav.edu/web/ciencia-razon-y-fe/puede-la-ciencia-ofrecer-una-explicacion-ultima-de-la-realidad#:~:text=Science%20cannot%20offer%20a%20complete,of%20the%20scientific%20method%20.1.

- How do we navigate between pragmatic and ultimate views of reality?

A balance between pragmatism and idealism can help you approach challenges with a balanced perspective and ensure your goals are both ambitious and attainable.[11]

6.4 The One and the Many

- How do we reconcile the apparent multiplicity of the world with notions of underlying unity?

The interplay between the one and the many is an age-old philosophical puzzle. Aurobindo tackles this by proposing that the many are simply the one expressing itself in various forms. This dialectic is not a contest but a dance — a cosmic choreography where the one becomes many to enjoy the play of existence.[12]

- What is the relationship between the absolute (Brahman) and the relative (world of experience)?

In Hinduism, the relationship between Brahman and the world of experience is that Brahman is the absolute reality and the world is a temporary, changing reality called Maya.[13]

6.5 The Role of Language and Concepts

- How adequate are our language and concepts for describing ultimate reality?

---

[11] https://chrysalis-services.in/pragmatism-vs-idealism/
[12] https://philosophy.institute/philosophy-of-sri-aurobindo/aurobindos-vision-divine-unity/
[13] https://iep.utm.edu/advaita-vedanta/

The limitedness of language only leads to partial incomprehension and partial misrepresentation of reality. Thus, it appears that reflection on language can doubtlessly serve to sharpen our awareness of the features of extra-linguistic reality and to make us notice what we possibly had not noticed before. However, that language can serve as an ultimate premise for inferring properties of the world, seems to be highly/naively questionable.[14]

- What are the limits of conceptual thinking in understanding the nature of reality?

Philosophers have thought more about the nature of thinking than about anything else. After Plato and Aristotle, philosophers' main concern was to promote good, that is, correct, thinking. Because correct thinking was achieved best in propositional statements, thinking became a matter of logic, and logic became a discipline dealing with the formulation of true predicative sentences. In the twentieth century, many philosophers expressed their dissatisfaction with this view. Some, such as Heidegger, have pointed to the ontological presuppositions of a logic that makes truth a matter of correspondence between predicative sentences and the reality of states of affairs. Other philosophers, such as Deleuze, have emphasized that the art of forming interesting philosophical problems cannot be reduced to the formulation of interrogative predicative sentences that can lead to a solution under the form of an affirmative predicative proposition.[15]

---

[14] *Udofia, Stephen. (2019). Language and the conception of reality. International Journal of Humanities and Innovation (IJHI). 2. 121-124. 10.33750/ijhi.v2i4.53.*
[15] *Bernet, Rudolf. (2014). The Limits of Conceptual Thinking. The Journal of Speculative Philosophy. 28. 219-241. 10.5325/jspecphil.28.3.0219.*

7. Scientific and Spiritual Paths to Understanding Reality

Both science and spiritual traditions offer paths to understanding the nature of reality, each with its own methods and insights:

7.1 The Scientific Method

- Empirical observation and experimentation

- Mathematical modeling

- Peer review and replication

- Technological applications

7.2 Spiritual Practices in Hindu Traditions

- Meditation and contemplation

- Yoga and other spiritual disciplines

- Study of sacred texts

- Guidance from spiritual teachers

7.3 Complementarity of Approaches

- How might scientific and spiritual approaches complement each other?

- What are the strengths and limitations of each approach?

8. Implications for Cosmology and Cosmic Origins

Our understanding of the nature of reality has profound implications for how we conceive of cosmic origins:

8.1 Quantum Cosmology

- Attempts to apply quantum mechanics to the early universe

- The Wheeler-DeWitt equation and the "wave function of the universe". The equation attempts to **mathematically combine the ideas of quantum mechanics and general relativity**, a step towards a theory of quantum gravity.

8.2 Creation in Hindu Thought

- Cycles of creation and dissolution

In Hindu cosmology, the cycle of creation and dissolution, or pralaya, is a repeated process of the universe being created, destroyed, and re-created.[16]

- The role of Maya in the manifestation of the universe

In Hinduism, the concept of Maya is that it is a powerful force that creates the illusion of the material world being real, while it is actually unreal. Maya is derived from the Sanskrit root "ma", which means to measure or delude.[17]

8.3 The Problem of Something from Nothing

The problem of something from nothing is the question of how something can exist when nothing is a complete absence of something. Explaining how something could come from nothing

---

[16] *https://en.wikipedia.org/wiki/Hindu_cosmology*
[17] *https://en.wikipedia.org/wiki/Maya_(religion)*

requires explaining the quantum state of the universe at the beginning of the Planck epoch.

- How quantum fluctuations might give rise to the universe?

- The concept of Brahman as the source of all manifestation

8.4 The Role of the Observer in Cosmic Origins

- Anthropic principles in cosmology: The anthropic principle is a cosmological hypothesis that the universe is structured to support life, and that life would not be possible if the fundamental physical constants of nature were slightly different. The principle is based on the idea that the existence of intelligent observers determines the universe's structure.

- The ultimate observer (Atman/Brahman) in Hindu thought

In Hinduism, the ultimate observer is Atman, the individual soul, which is part of Brahman, the ultimate reality.[18]

9. Practical and Existential Implications

How do these views of reality affect our lived experience and our approach to life?

9.1 Ethical Considerations

- How might our understanding of reality influence our ethical frameworks?

Our understanding of reality can influence our ethical frameworks by shaping our beliefs and values, which are the basis for our ethical decision-making. For example, our

---

[18] *https://en.wikipedia.org/wiki/Brahman*

perceptions and experiences can lead us to believe things are true, even when they are not.[19]

- The concept of interconnectedness and its ethical implications

The concept of interconnectedness can widen the range and depth of ethical and moral concerns. It can help people see that the interests of others are their own interests, and that their suffering is their own suffering.[20]

9.2 Meaning and Purpose

- How do different views of reality relate to questions of meaning and purpose?

Different views of reality can relate to questions of meaning and purpose in a number of ways.[21]

- Navigating between ultimate and conventional views of reality

The theory of two truths is a framework used in Buddhism to navigate between the conventional and ultimate views of reality.[22]

9.3 Approaches to Knowledge

- Balancing scientific and contemplative approaches to understanding reality

Due to ignorance about the nature of sense data, most people process perceptual data with the erroneous view that we humans have a separate internal essence that interacts directly

---

[19] https://xmonks.com/ethical-decision-making/
[20] https://philarchive.org/archive/PEETEO-9
[21] https://en.wikipedia.org/wiki/Reality
[22] https://tricycle.org/beginners/buddhism/two-truths/

with and somehow possesses objects in the material world. To emphasize again why this view is erroneous, con- sider that even if we were to directly grasp external material phenomena, instead of just perceptual data, recognition of objects would still be fundamentally arbitrary and dependent on the individual's conceptual schema, which is itself inherently mutable. One person's biohazard trash heap may be another person's treasured archaeological record – but only after that other person has received archaeological training.[23]

- The role of direct experience in understanding reality

Direct experience, or immediate experience, is the idea that knowledge and skills are gained through the senses and cannot be fully described with words. Some say that direct experience is the mind coming into direct contact with the object of knowledge, rather than just the experience of the senses.[24]

9.4 Personal Transformation

- How might realizing the nature of reality (whether through scientific or spiritual means) transform our lives?

Realizing the nature of reality can transform our lives by helping us to understand the complexity of the universe.Science can help us understand the balance and complexity of the universe, which can lead to introspection and a sense of reverence for the universe. Science and spirituality both share the idea that our

---

[23] *Wright, M.J., Sanguinetti, J.L., Young, S. et al. Uniting Contemplative Theory and Scientific Investigation: Toward a Comprehensive Model of the Mind. Mindfulness* **14***, 1088–1101 (2023). https://doi.org/10.1007/s12671-023-02101-y*
[24] *https://www.diamondapproach.org/glossary/refinery_phrases/direct-experience*

physical senses can deceive us, and that there is a deeper reality to be discovered.[25]

- The concept of liberation or enlightenment in relation to understanding reality

Liberation and enlightenment are concepts that relate to understanding reality and achieving a state of supreme wisdom.[26]

10. Conclusion: Towards a More Comprehensive View of Reality

As we've explored in this chapter, both quantum mechanics and the Hindu concept of Maya challenge our conventional notions of reality. They suggest a world that is more fluid, interconnected, and mysterious than our everyday experience might indicate. While these perspectives arise from very different contexts and methodologies, they both point towards a reality that transcends simple materialism or naive realism.

The parallels between these views are striking: both suggest a reality beyond appearances, emphasize the role of the observer, point to fundamental interconnectedness, and imply limits to our ordinary ways of knowing. Yet, they also differ in significant ways, particularly in their methodologies and the ultimate conclusions they draw about the nature of reality.

As we continue our exploration of cosmic origins, bridging the Big Bang and Brahman, this multifaceted understanding of reality will be crucial. It allows us to appreciate both the scientific quest

---

[25] *https://philosophy.fsu.edu/undergraduate-study/why-philosophy/What-is-Philosophy*
[26] *https://en.wikipedia.org/wiki/Moksha*

to understand the physical universe and the spiritual search for ultimate truth. It challenges us to hold multiple perspectives simultaneously, to navigate between pragmatic and ultimate views of reality, and to remain open to the profound mystery at the heart of existence.

In the next chapter, we will delve into another fundamental aspect of reality that bridges our scientific and spiritual inquiries: the nature of consciousness and its role in the cosmos.

6. Consciousness and the Cosmos

Introduction: The Enigma of Consciousness

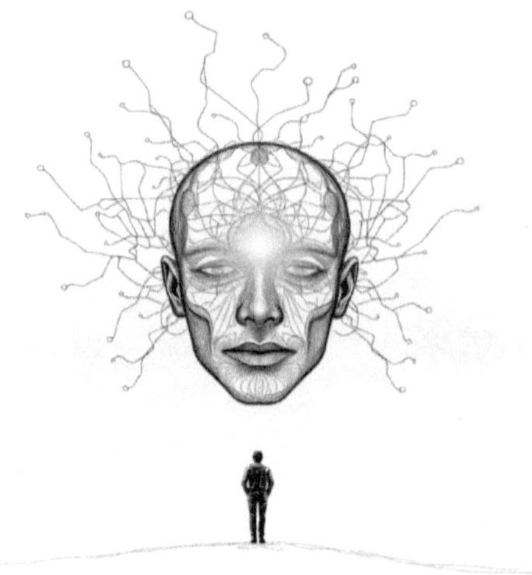

Consciousness, the subjective experience of being aware, stands as one of the most profound mysteries in our understanding of reality. It is the lens through which we perceive the universe, yet its nature and relationship to the physical world remain deeply puzzling. As we continue our exploration of cosmic origins, bridging the Big Bang and Brahman, we find that consciousness plays a crucial role in both scientific and spiritual perspectives on the cosmos.

In modern science, consciousness has emerged as a frontier of inquiry, challenging our understanding of the brain, mind, and the nature of reality itself. From the role of the observer in quantum mechanics to the hard problem of consciousness in philosophy of mind, questions about awareness and experience permeate cutting-edge research.

In Hindu philosophy, particularly in schools like Advaita Vedanta, consciousness (often termed Chit) is not merely a product of material processes but is considered fundamental to the nature of reality. The concept of Brahman, as we've explored in previous chapters, is often described as Sat-Chit-Ananda (Being-Consciousness-Bliss), placing consciousness at the very heart of ultimate reality.

In this chapter, we will delve into the nature of consciousness from both scientific and Hindu philosophical perspectives. We will explore how these different approaches conceptualize consciousness, its relationship to the physical world, and its potential role in the cosmos at large. By examining these questions through multiple lenses, we aim to develop a richer, more nuanced understanding of consciousness and its place in our conception of cosmic origins.

2. Scientific Perspectives on Consciousness

The scientific study of consciousness has gained significant momentum in recent decades, drawing insights from neuroscience, psychology, physics, and philosophy. Let's explore key aspects of the scientific approach to consciousness:

2.1 Neuroscientific Approaches

Neuroscience seeks to understand consciousness by studying the brain and nervous system:

- **Neural Correlates of Consciousness (NCCs)**: Identifying brain activity patterns associated with conscious experiences.

- **Global Workspace Theory**: Proposes that consciousness arises from the global broadcast of information in the brain.

- **Integrated Information Theory**: Suggests that consciousness is intrinsic to certain physical systems and can be quantified.

2.2 Cognitive and Psychological Approaches

Cognitive science and psychology examine consciousness through mental processes and behavior:

- **Attention and Awareness**: Studying the relationship between attention mechanisms and conscious perception.

- **Unconscious Processing**: Exploring the vast realm of mental processes that occur outside of awareness.

- **Altered States of Consciousness**: Investigating experiences like meditation, psychedelic states, and near-death experiences.

2.3 Quantum Approaches to Consciousness

Some researchers have proposed that quantum mechanics may play a role in consciousness:

- **Orchestrated Objective Reduction (Orch-OR)**: Penrose and Hameroff's theory suggesting quantum processes in microtubules give rise to consciousness.

- **Quantum Mind Theories**: Various proposals that quantum effects in the brain could explain features of consciousness like free will or non-locality.

2.4 Philosophical Approaches

Philosophy of mind grapples with fundamental questions about the nature of consciousness:

- The Hard Problem of Consciousness: David Chalmers' formulation of the challenge of explaining why we have subjective experiences.

- Dualism vs. Materialism: Debates about whether consciousness is separate from or emergent from physical processes.

- Panpsychism: The view that consciousness is a fundamental feature of the universe, present to some degree in all matter.

2.5 Artificial Intelligence and Consciousness

The development of AI raises new questions about the nature of consciousness:

- Can machines be conscious? What would be required for artificial consciousness?

- The Chinese Room Argument and other challenges to machine consciousness.

The Chinese Room Argument is a thought experiment that challenges the idea that computers can understand language or think. The argument is based on the idea that computers can only manipulate symbols according to syntactic rules, and cannot understand the meaning of human input. The thought experiment imagines someone who is only able to speak English sitting in a room and following instructions to manipulate Chinese characters. To those outside the room, it would appear as if the person in the room understands Chinese, but in reality they do not.[27]

- Implications of potential artificial consciousness for our understanding of mind and reality.

3. Consciousness in Hindu Philosophy

Hindu philosophy, particularly in its Vedantic forms, places consciousness at the center of its understanding of reality.

Vedanta accepts five means of knowledge, namely, Pratyaksha, Anuman, Upamana, Arthapatti and Shastra or Shabda. The explanation of these is given below:

- **Pratyaksha**: Perception, or direct sensory experience

- **Anumana**: Inference, or reasoning based on perception

- **Upamana**: Comparison, or analogy

- **Arthapatti**: Presumption, or postulation to explain unobservable phenomena

- **Shastra/Shabda**: Verbal testimony, or credible source

The first four are used to gain knowledge of the world as we perceive. The last means is to know Brahman.

The first four depend on perception (pratyaksha) as the basic means. Perception has the limitations of the perceiver, the instruments and the perceived object. Thus it cannot give any holistic knowledge.

---

[27] *https://plato.stanford.edu/archIves/spr2010/entries/chinese-room/*

Hence even science which uses these four means always finds a limit to whatever they understand. They always arrive at some subtle effect rather than the ultimate cause of the world.

Shastra (Scriptures) praman (means of knowledge) which is Anadi (without beginning), Ananta (endless) and Apaurusheya (not of a human origin) alone remains the means to know about the ultimate cause of this world and its nature.

The knowledge happens in Buddhi (Intellect). When we use the above four means to know the world the appropriate thoughts takes place in the buddhi which we call knowledge.

When we use the Shastra (Scriptures) as means to know Brahman. It uses the technique of neti neti (negation- not this not this). When all the qualities of world are negated what remains in the buddhi is an image of Brahman. Like in a mirror when all objects are negated what remains is an image of space. This image is called Akhandakar vritti (constant awareness). Akhandakara vritti is a concept in Vedanta that describes the constant awareness of being the light of consciousness in every objective knowledge. It means that even though the vrittis (awareness) of objects are broken, there is a continuous flow of thoughts and reflection of the light of consciousness.

This is a theoretical understanding of Brahman. Shastra says this Brahman is not different from our own self which is experienced as thoughtless 'I' (as in deep sleep). To constantly assert this Brahman as our self is the practice is nididhyasana (vedantic meditation) which ultimately culminates in experience which is called Aparokshanubuti (self realisation).

Let's explore key concepts related to consciousness in Hindu thought:

3.1 Atman and Brahman

- Atman: The individual self or consciousness, often considered identical with Brahman.

- Brahman: The ultimate reality, often described as pure consciousness (Chit).

- The Mahavakya(Great Saying) "Tat Tvam Asi" (That Thou Art): Expressing the identity of individual consciousness with universal consciousness. The Mahāvākyas are "The Great Sayings" of the Upanishads, as characterized by the Advaita school of Vedanta with mahā meaning great and vākya, a sentence.

3.2 Levels of Consciousness

Many Hindu texts describe different states or levels of consciousness:

- **Waking (Jagrat)**: Ordinary waking consciousness.

- **Dreaming (Swapna)**: The consciousness experienced in dreams.

- **Deep Sleep (Sushupti)**: A state of potential consciousness without objects.

- **Turiya**: The fourth state, transcending the other three, often equated with pure consciousness.

3.3 Consciousness as Fundamental

- In Advaita Vedanta, consciousness is not an emergent property but the fundamental reality.

- The world of objects is seen as appearances within consciousness rather than independent realities.

3.4 The Witness Consciousness

- The concept of Sakshi (witness) consciousness that observes but does not engage with experiences.

- The practice of cultivating witness awareness in meditation and daily life.

3.5 Consciousness and Maya

- The role of consciousness in the creation and perception of the illusory world (Maya).

- How ignorance (Avidya) veils the true nature of consciousness as Brahman.

3.6 Yoga and the Science of Consciousness

- Patanjali's Yoga Sutras as a systematic approach to understanding and exploring consciousness.

- The goal of Yoga as the stilling of mental modifications to reveal pure consciousness.

4. Comparative Analysis: Scientific and Hindu Views on Consciousness

Despite their different methodologies and contexts, scientific and Hindu approaches to consciousness offer some intriguing parallels and contrasts:

4.1 The Primacy of Consciousness

- Some interpretations of quantum mechanics suggest a fundamental role for consciousness in reality.

The von Neumann–Wigner interpretation

This interpretation suggests that consciousness is necessary for the completion of quantum measurement. Eugene Wigner argued that human consciousness is critical for the collapse of the wave function. [28]

- Hindu philosophy, particularly Advaita Vedanta, posits consciousness as the ultimate reality.

4.2 The Observer Effect

- The role of the observer in quantum mechanics has parallels with the concept of witness consciousness in Hindu thought.

The act of measuring a quantum particle affects its properties because the measurement requires interacting with the particle. This is known as the observer effect.[29]

- Both perspectives challenge the notion of an objective reality independent of consciousness.

4.3 Unity of Consciousness

- Some scientific theories propose an integrated or unified nature of consciousness (e.g., Integrated Information Theory).

This Integrated Information Theory (IIT)suggests that consciousness is a type of information that can be measured

---

[28] https://en.wikipedia.org/wiki/Von_Neumann%E2%80%93Wigner_interpretatio n

[29] https://en.wikipedia.org/wiki/Observer_(quantum_physics)

mathematically. It proposes that consciousness is dependent on a system's ability to integrate information, and that consciousness increases as the number of available states in a system increases. IIT aims to explain why some physical systems, like the human brain, are conscious, and to determine if other systems are conscious.[30]

- Hindu philosophy often describes a fundamental unity of all consciousness in Brahman.

4.4 Altered States of Consciousness

- Scientific interest in meditation, psychedelics, and other altered states aligns with Hindu explorations of different levels of consciousness.

- Both approaches recognize the potential for expanded or transcendent states of awareness.

4.5 The Hard Problem of Consciousness

- The scientific struggle with the hard problem of consciousness resonates with the Hindu recognition of consciousness as ultimately mysterious and self-revealing.

The hard problem of consciousness is the scientific challenge of explaining why a physical state is conscious, and why a subject has a conscious experience. It's a difficult problem because it goes beyond the usual methods of science, which explain how things function, change over time, and are put together. [31]

4.6 Consciousness and Reality

---

[30] *https://iep.utm.edu/integrated-information-theory-of-consciousness/*
[31] *https://iep.utm.edu/hard-problem-of-conciousness/*

- Some scientific perspectives (e.g., Wheeler's participatory universe) suggest a role for consciousness in shaping reality.

Wheeler asserted, that the "laws" of the functioning of the Universe (physics) make man's participation in the flow of events – in the observable material reality – a given. And if that is true, then, it follows, that that participation leads to more "creation" (actions by man) which, as Wheeler put it, is new "information" added to the world (the observable reality) which gives rise to (more) physics (more material effects in the Universe).[32]

- Hindu philosophy often describes the manifest world as a play of consciousness (Lila).

5. Consciousness and Cosmic Origins

How might our understanding of consciousness inform our views on the origin and nature of the cosmos?

5.1 Consciousness in Cosmology

- Anthropic principle in cosmology and their relation to consciousness. This principle is discussed later in chapter 8.

The anthropic principle is a methodological principle that suggests that the universe's parameters must be such that life is possible. It's based on the idea that life would not have been able to exist if the fundamental constants of nature had been slightly different.[33]

---

[32] https://futurism.com/john-wheelers-participatory-universe
[33] https://en.unav.edu/web/ciencia-razon-y-fe/john-barrow-y-el-principio-cosmologico-antropico

- The role of observation in cosmological models and the implications for a conscious universe.

5.2 Panpsychist Cosmologies

- Proposals that consciousness might be a fundamental feature of the universe, present from its very beginning.

- How such views might bridge scientific and Hindu perspectives on cosmic origins.

5.3 Creation as an Act of Consciousness

- Hindu cosmologies often describe creation as a manifestation or dream of cosmic consciousness.

- Parallels with scientific notions of reality as information or computation.

5.4 The Evolution of Consciousness

- Scientific perspectives on the emergence and evolution of consciousness in the cosmos.

- Hindu views on the unfoldment of consciousness through different forms and levels of manifestation.

6. Philosophical Implications

The study of consciousness raises profound philosophical questions that bridge scientific and spiritual inquiries:

6.1 The Nature of Self

- How does our understanding of consciousness affect our concept of self?

- The Buddhist doctrine of Anatta (no-self) and its relevance to both scientific and Hindu views.

6.2 Free Will and Determinism

- How does consciousness relate to questions of free will?

- Reconciling apparent free will with both physical laws and the concept of a unified consciousness.

6.3 The Mind-Body Problem

- Various approaches to understanding the relationship between consciousness and physical reality.

- Parallels between scientific theories of mind-body relationship and Hindu concepts of subtle bodies.

6.4 Ethics and Consciousness

- How might our understanding of consciousness inform our ethical frameworks?

- The concept of expanded circles of compassion based on the unity of consciousness.

6.5 The Limits of Knowledge

- Can consciousness fully understand itself?

- The role of direct experience versus conceptual knowledge in understanding consciousness.

7. Practical Applications and Implications

How do these perspectives on consciousness translate into practical approaches and implications?

7.1 Meditation and Consciousness Exploration

- Scientific studies on the effects of meditation on the brain and consciousness.

- Hindu practices for exploring and expanding consciousness.

7.2 Consciousness in Healthcare

- The role of consciousness in healing and wellbeing.

- Integrating consciousness-based approaches with modern medicine.

7.3 Education and Consciousness

- Implications for educational approaches based on our understanding of consciousness.

- Cultivating awareness and metacognition in learning processes.

7.4 Technology and Consciousness

- The development of brain-computer interfaces and their implications for our understanding of consciousness.

- Ethical considerations in the development of AI and potential artificial consciousness.

7.5 Environmental Perspectives

- How our view of consciousness might inform our relationship with the environment.

- The concept of Earth as a conscious system (e.g., Gaia hypothesis) and its resonance with Hindu views.

8. Consciousness and the Future of Cosmic Exploration

As we look to the future of both scientific and spiritual inquiry, consciousness plays a central role:

8.1 The Final Frontier

- Consciousness as the next great frontier in scientific exploration.

- The role of inner exploration in complementing outer space exploration.

8.2 Integrating Perspectives

- The potential for a more integrated approach to studying consciousness that bridges scientific and contemplative methods.

- Challenges and opportunities in creating dialogue between different approaches to consciousness.

8.3 Consciousness and Human Potential

- Exploring the furthest reaches of human consciousness and its implications for our understanding of reality.

- The concept of enlightenment or self-realization in light of scientific understandings of consciousness.

8.4 Cosmic Consciousness

- Speculations on the possibility of cosmic-scale consciousness or universal awareness.

- How such concepts might inform our understanding of the meaning and purpose of the cosmos.

9. Conclusion: Consciousness as a Bridge

As we've explored in this chapter, consciousness serves as a crucial bridge between scientific and spiritual approaches to understanding the cosmos. Both perspectives recognize consciousness as a fundamental mystery, central to our experience of reality and potentially to the nature of reality itself.

The scientific study of consciousness, while still in its early stages, is revealing the profound complexity of awareness and its intimate relationship with the physical world. From the role of the observer in quantum mechanics to the global integration of information in the brain, science is uncovering layers of subtlety in how consciousness operates and relates to the world.

Hindu philosophy, with its millennia-long exploration of consciousness, offers profound insights into the nature of awareness, its potential states and dimensions, and its relationship to ultimate reality. The concept of consciousness as fundamental, rather than emergent, challenges materialist assumptions and opens up new ways of conceiving the cosmos.

As we continue our exploration of cosmic origins, bridging the Big Bang and Brahman, this multifaceted understanding of consciousness will be crucial. It allows us to appreciate both the empirical study of the cosmos and the direct exploration of consciousness itself as valid and complementary approaches to understanding reality.

The convergence of scientific and contemplative approaches to consciousness holds immense potential for deepening our understanding of ourselves and the universe. As we move forward, the study of consciousness may well prove to be the key to unlocking the deepest mysteries of existence, offering a unified vision of the cosmos that embraces both its physical and experiential dimensions.

7. Evolution and Cycles of Creation

Introduction: The Unfolding Cosmos

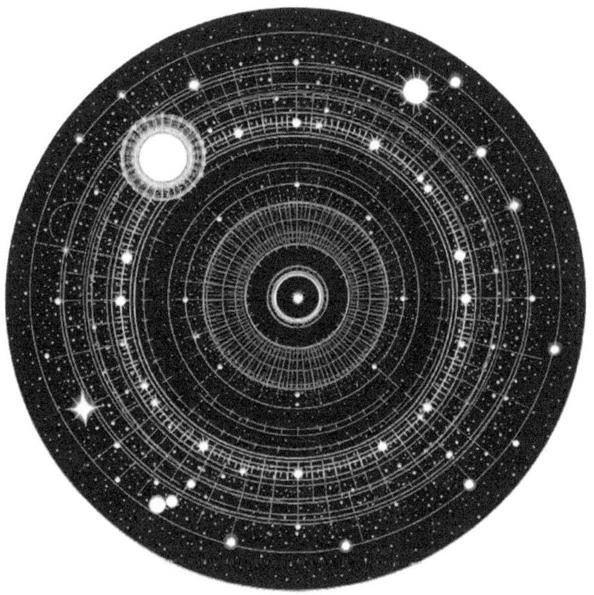

As we continue our exploration of cosmic origins, bridging the scientific perspective of the Big Bang with the Hindu concept of Brahman, we encounter two profound and seemingly contrasting views of cosmic development: the linear evolution described by modern cosmology and the cyclical creation concepts found in Hindu philosophy. Both perspectives offer deep insights into the nature of the cosmos and its development over time, challenging us to expand our understanding of the universe's history and future.

Modern cosmology, building on the Big Bang theory, presents a picture of cosmic evolution that spans nearly 14 billion years. From the initial singularity through the formation of fundamental particles, atoms, stars, and galaxies, to the complex universe we observe today, this scientific narrative describes a linear progression of increasing complexity and structure.

In contrast, many schools of Hindu thought conceive of cosmic development in vast cycles. These cycles, often described in terms of the days and nights of Brahma or the concept of Yugas, suggest a rhythmic pattern of creation, preservation, and dissolution that repeats on an epic scale.

In this chapter, we will explore both the scientific understanding of cosmic evolution and the Hindu concepts of cyclical creation. We will examine the evidence and reasoning behind each perspective, consider their implications for our understanding of time and change, and explore potential areas of convergence between these seemingly disparate worldviews.

By engaging with both scientific and Hindu perspectives on cosmic development, we aim to develop a richer, more nuanced understanding of the universe's past, present, and potential future. This exploration may offer new insights into the nature of time, change, and the fundamental patterns that govern the cosmos.

2. Cosmic Evolution in Modern Science

Modern cosmology and astrophysics have developed a detailed narrative of cosmic evolution, supported by observational evidence and theoretical models. Let's explore the key stages and concepts in this scientific understanding of the universe's development:

2.1 The Big Bang and Early Universe

The story of cosmic evolution begins with the Big Bang, the theoretical starting point of our observable universe:

- **Planck Era** (before 10^{-43} seconds): The earliest conceivable time, where our current physics breaks down.

- **Grand Unification Era** (10^{-43} to 10^{-36} seconds): All forces except gravity are unified.

- **Inflationary Era** (10^{-36} to 10^{-32} seconds): A period of exponential expansion of the early universe.

- **Electroweak Era** (10^{-32} to 10^{-12} seconds): The strong force separates, leaving electroweak force.

- **Quark Era** (10^{-12} to 10^{-6} seconds): Quarks and gluons form a quark-gluon plasma.

- **Hadron Era** (10^{-6} to 1 second): Quarks combine to form hadrons like protons and neutrons.

2.2 Nucleosynthesis and the Cosmic Dark Ages

- **Primordial Nucleosynthesis** (3 minutes to 20 minutes): Formation of light elements (mainly hydrogen and helium).

- **Recombination** (380,000 years): The universe cools enough for atoms to form, releasing the cosmic microwave background radiation.

- **Cosmic Dark Ages** (380,000 years to 150 million years): A period before the first stars, when the universe was dark and mostly uniform.

2.3 Structure Formation

- **First Stars** (150-200 million years): The first generation of stars (Population III) form.

- **Reionization** (150 million to 1 billion years): The first stars and galaxies reionize the neutral hydrogen in the universe.

- **Galaxy Formation** (1 billion years onwards): Galaxies begin to form and evolve.

- **Large Scale Structure**: Over time, galaxies organize into clusters, filaments, and voids, creating the cosmic web.

2.4 The Evolution of Stars and Chemical Enrichment

- **Stellar Lifecycles**: The birth, life, and death of stars of various masses.

- **Nucleosynthesis in Stars**: The creation of heavier elements within stars and in supernovae.

- **Chemical Evolution**: The gradual enrichment of the universe with heavier elements.

2.5 Formation of Planetary Systems

- **Protoplanetary Disks**: The formation of disks around young stars.

- **Planet Formation**: The aggregation of dust and gas into planets.

- **The Search for Exoplanets**: Our growing understanding of the diversity of planetary systems.

2.6 The Future of Cosmic Evolution

Scientific models also make predictions about the future evolution of the universe:

- **The Fate of the Solar System**: The future of our sun and planets.

- **Galaxy Mergers**: The predicted collision of the Milky Way and Andromeda galaxies.

- **The Far Future**: Scenarios like the heat death of the universe or the Big Rip, depending on the nature of dark energy.

3. Cycles of Creation in Hindu Thought

Hindu cosmology presents a richly detailed vision of cosmic cycles, offering a perspective on universal rhythms that span vast periods of time. Let's explore some key concepts in this cyclical view of creation:

3.1 The Day and Night of Brahma

One of the most fundamental cycles in Hindu cosmology is the alternation between the day and night of Brahma:

- **Brahma's Day** (Kalpa): A period of cosmic manifestation lasting 4.32 billion years.

- **Brahma's Night**: An equal period of cosmic dissolution.

- **The life of Brahma**: 100 Brahma years, each consisting of 360 days and nights.

3.2 The Four Yugas

Within each day of Brahma, there are smaller cycles known as Yugas:

- **Satya Yuga** (Golden Age): A time of virtue and spiritual purity.

- **Treta Yuga** (Silver Age): A period of decline from the golden age.

- **Dvapara Yuga** (Bronze Age): Further decline in virtue and spirituality.

- **Kali Yuga** (Iron Age): The age of conflict and spiritual darkness.

These four Yugas together form a Maha Yuga, and 1000 Maha Yugas make one day of Brahma.

3.3 Manvantaras

Each Kalpa is divided into 14 Manvantaras, periods presided over by different Manus (progenitors of humanity):

- Each Manvantara lasts about 306.72 million years.

- They represent different cycles of human civilization and cosmic administration.

3.4 Pralaya and Mahapralaya

Hindu cosmology describes different levels of cosmic dissolution:

- **Pralaya**: A partial dissolution at the end of each Manvantara.

- **Mahapralaya**: The great dissolution at the end of Brahma's life, where the entire manifest universe is reabsorbed into the unmanifest.

3.5 The Breath of Vishnu

Some traditions describe the cycles of creation and dissolution as the breath of Vishnu:

- **Exhalation**: The period of cosmic manifestation.

- **Inhalation**: The period of cosmic dissolution.

3.6 Eternal Cycles

A key feature of Hindu cosmology is the concept of eternal cycles:

- No absolute beginning or end to the cycles.

- Each cycle is part of an endless sequence of creation, preservation, and dissolution.

4. Comparative Analysis: Linear Evolution and Cyclical Creation

While the scientific view of cosmic evolution and the Hindu concept of cyclical creation may seem fundamentally different, there are intriguing areas of convergence and complementarity:

4.1 Time Scales

- Both perspectives deal with vast time scales, far beyond human experience.

- The scientific age of the universe (13.8 billion years) is comparable to some of the time scales in Hindu cosmology.

4.2 Patterns of Development

- Science describes patterns of stellar and galactic evolution that repeat throughout the universe.

- Hindu cycles suggest repeating patterns of cosmic and civilizational development.

4.3 Entropy and Cosmic Cycles

- The scientific concept of entropy and the heat death of the universe has some parallels with the idea of cosmic dissolution in Hindu thought.

- Some speculative scientific models propose cyclic universes, resonating with Hindu cyclical cosmology.

These models propose cyclic universes, which are models that suggest the universe goes through endless cycles of expansion and contraction.

Conformal cyclic cosmology

A theory based on general relativity that proposes the universe goes through repeated cycles of expansion, each starting with a "big bang" and ending in a stage of accelerated expansion. This theory was proposed by Roger Penrose, a 2020 Nobel Prize-winning physicist. [34]

4.4 Complexity and Simplicity

- Both perspectives describe alternations between states of greater and lesser complexity.

- The scientific view of the universe evolving from a simple early state to complex structures, and potentially back to a simple state (e.g., heat death), has parallels with the Hindu concept of manifestation and dissolution.

---

[34] *https://en.wikipedia.org/wiki/Cyclic_model*

4.5 The Role of Consciousness

- Some interpretations of quantum mechanics suggest a role for consciousness in the evolution of the universe.

- Hindu cosmology often describes cosmic cycles as the play (lila) or dream of cosmic consciousness.

5. Philosophical Implications

The study of cosmic evolution and cycles raises profound philosophical questions:

5.1 The Nature of Time

- Is time fundamentally linear or cyclical?

- How do we reconcile our subjective experience of time with cosmic time scales?

5.2 Progress and Purpose

- Does cosmic evolution imply progress or purpose?

- How do cyclical views of time affect our understanding of progress?

5.3 Determinism and Free Will

- How much of cosmic evolution is deterministic, and is there room for free will or chance?

- How do Hindu concepts of karma and divine play relate to questions of determinism?

5.4 The Role of Life and Consciousness

- What is the place of life and consciousness in cosmic evolution?

- Are we passive observers of cosmic processes, or active participants?

5.5 Ultimate Origins

- Can we meaningfully speak of an ultimate origin of the cosmos?

- How do linear and cyclical views address the question of first causes?

6. Implications for Cosmology and Cosmic Origins

Our understanding of evolution and cycles has profound implications for how we conceive of cosmic origins:

6.1 The Beginning of Time

- The Big Bang theory suggests a beginning of time, but some cyclic models propose no absolute beginning.

- Hindu cosmology often describes cycles without beginning or end, challenging the notion of an absolute origin.

6.2 The Fate of the Universe

- Scientific models make various predictions about the universe's fate (e.g., heat death, big rip, big crunch).

- Hindu cyclical models suggest ongoing cycles of creation and dissolution.

6.3 Multiverse Theories

- Some scientific theories propose multiple universes or repeating universes.

- These have interesting parallels with Hindu concepts of multiple world systems or repeating creations.

6.4 The Anthropic Principle

- The fine-tuning of universal constants raises questions about cosmic evolution and the emergence of life.

- Cyclical models offer a different perspective on the apparent fine-tuning of the universe.

7. Evolution, Cycles, and Human Understanding

How do these perspectives on cosmic development inform our understanding of ourselves and our place in the universe?

7.1 Cosmic Perspective

- Both scientific and Hindu views encourage us to consider our place in vast cosmic processes.

- This cosmic perspective can profoundly affect our worldview and values.

7.2 The Nature of Change

- Understanding cosmic evolution and cycles can inform our approach to change in our own lives and societies.

- It raises questions about the nature of progress and the cycles of rise and fall in civilizations.

7.3 Environmental Implications

- Our understanding of cosmic processes can inform our approach to environmental stewardship.

- Long-term thinking inspired by cosmic time scales may influence our decision-making.

7.4 Spiritual and Existential Implications

- How do different models of cosmic development affect our search for meaning and purpose?

- What are the implications for spiritual practices and beliefs?

8. Integrating Perspectives

As we seek to bridge scientific and Hindu perspectives on cosmic origins, several approaches to integration emerge:

8.1 Complementary Descriptions

- Viewing scientific and Hindu models as complementary descriptions of reality, each valuable in its own context.

- Exploring how these different perspectives might inform and enrich each other.

8.2 Metaphorical Interpretation

- Interpreting cyclical models metaphorically, as describing psychological or spiritual processes rather than literal cosmic events.

- Examining how such interpretations might complement scientific understanding.

8.3 Speculative Synthesis

- Exploring speculative scientific models (e.g., cyclic universes, emergent time) that might align with aspects of Hindu cosmology.

- Considering how Hindu insights might inspire new directions in scientific research.

8.4 Epistemological Reflection

- Reflecting on the methods and limits of both scientific and spiritual approaches to understanding cosmic processes.

- Considering how different ways of knowing might be integrated for a more comprehensive understanding.

9. Conclusion: Rhythms of the Cosmos

As we've explored in this chapter, both modern science and Hindu philosophy offer profound insights into the development of the cosmos. The scientific narrative of cosmic evolution presents a majestic unfolding of increasing complexity and structure over billions of years. Hindu concepts of cosmic cycles offer a vision of grand rhythms of creation, preservation, and dissolution repeating on unimaginable time scales.

While these perspectives may seem at odds, they both encourage us to expand our vision beyond the limited scope of human experience. Both scientific and Hindu views reveal a cosmos of awe-inspiring scale and complexity, inviting us to contemplate our place within these grand processes.

The apparent tension between linear evolution and cyclical creation may ultimately prove fertile ground for deeper understanding. It challenges us to think more deeply about the nature of time, change, and the fundamental patterns that govern the cosmos. As we continue to probe the mysteries of the universe, both through scientific inquiry and contemplative insight, we may find new ways to harmonize these perspectives, leading to a richer, more nuanced understanding of our cosmic origins and destiny.

In the next chapter, we will explore another fundamental aspect of cosmic origins: the fine-tuning of the universe and its relationship to concepts of divine plan or cosmic order in both scientific and Hindu thought.

8. The Anthropic Principle and Divine Plan

1. Introduction: A Universe Finely Tuned for Life

As we delve deeper into our exploration of cosmic origins, bridging the scientific perspective of the Big Bang with the Hindu concept of Brahman, we encounter a profound and perplexing question: Why does the universe appear to be so exquisitely fine-tuned for the existence of life? This question leads us to consider two intriguing concepts: the anthropic principle in modern cosmology and the notion of divine plan or cosmic order (often referred to as Rta (Sanskrit ऋत) in Hindu thought. In Hindu philosophy, Ṛta is the principle of natural order that governs the universe and everything in it. It is a Vedic legal concept that can be translated as "natural law". The word Ṛta comes from the root ri, which means "to move".[35]

The anthropic principle, in its various forms, suggests that the observed properties of the universe are constrained by the necessity of the existence of life and consciousness. It arises from the observation that many fundamental constants and laws of physics seem to be "just right" for the emergence of complexity and life. If these constants were even slightly different, the universe as we know it, including life, could not exist.

In Hindu philosophy, particularly in Vedic and later Hindu thought, the concept of Rta (Sanskrit ऋत) represents the principle of cosmic order. It suggests an underlying harmony and purpose in the universe, a divine plan that governs both natural and moral laws. This concept resonates with the idea of a universe designed or fine-tuned for a purpose.

In this chapter, we will explore both the anthropic principle and the Hindu concept of cosmic order. We will examine the scientific evidence for fine-tuning, consider various interpretations of the anthropic principle, and delve into Hindu perspectives on cosmic order and purpose. By engaging with both scientific and Hindu viewpoints, we aim to develop a richer, more nuanced understanding of the apparent order and purposefulness we observe in the cosmos.

This exploration may offer new insights into questions of purpose, design, and the place of consciousness in the universe,

---

[35] *https://www.linkedin.com/pulse/rta-dharma-ancient-indian-concept-law-justice-krishna-das/*

challenging us to think deeply about our cosmic origins and the nature of reality itself.

2. The Anthropic Principle in Modern Cosmology

The anthropic principle has emerged as a significant topic in modern cosmology and philosophy of science. Let's explore its key aspects and implications:

2.1 Fine-Tuning in the Universe

The concept of fine-tuning refers to the observation that many fundamental constants and laws of physics seem precisely calibrated to allow for the existence of life:

- **The Cosmological Constant**: The energy density of empty space is fine-tuned to within 1 part in 10^{120}.

- **The Strong Nuclear Force**: If it were slightly weaker, only hydrogen would exist; if slightly stronger, no stars would form.

- **The Weak Nuclear Force**: Its strength affects the production of carbon and oxygen in stars.

- **The Electromagnetic Force**: Its strength determines chemical bonding and the stability of atoms.

- **The Ratio of Proton to Electron Mass**: Affects the stability of atoms and the formation of molecules.

2.2 Types of Anthropic Principles

Several versions of the anthropic principle have been proposed:

2.2.1 Weak Anthropic Principle (WAP)

- States that the observed values of physical and cosmological quantities are restricted by the requirement that they allow the existence of observers to measure them.

- Often interpreted as a selection effect: we can only observe universes compatible with our existence.

2.2.2 Strong Anthropic Principle (SAP)

- Suggests that the universe must have properties that allow life to develop within it at some stage in its history.

- Sometimes interpreted as implying that the universe is in some sense compelled to eventually create observers.

2.2.3 Final Anthropic Principle (FAP)

- Proposes that intelligent information-processing must come into existence in the universe, and, once it comes into existence, it will never die out.

2.2.4 Participatory Anthropic Principle (PAP)

- Suggests that observers are necessary to bring the universe into existence, drawing on interpretations of quantum mechanics that emphasize the role of observation.

2.3 Multiverse Theories and the Anthropic Principle

Some scientists propose multiverse theories to explain fine-tuning:

- Inflationary Multiverse: Suggests that our universe is one of many "bubble universes" with different physical constants.

- String Theory Landscape: Proposes a vast number of possible vacuum states, each representing a different universe with different physical laws.

- Many-Worlds Interpretation of Quantum Mechanics: Suggests that all possible alternate histories and futures are real, each representing an actual world or parallel universe.

In these scenarios, the anthropic principle becomes a selection effect: we naturally find ourselves in a universe compatible with our existence.

2.4 Criticisms and Controversies

The anthropic principle has been subject to various criticisms:

- Tautology: Some argue that the weak anthropic principle is tautological and lacks explanatory power.

- Anthropocentrism: Critics suggest it places undue emphasis on human existence.

- Lack of Predictive Power: Unlike other scientific principles, it doesn't make testable predictions.

- Alternative Explanations: Some argue that future scientific discoveries might explain apparent fine-tuning without resorting to anthropic reasoning.

3. Cosmic Order (Rta) in Hindu Philosophy

The concept of Rta in Hindu philosophy offers a different perspective on cosmic order and purpose. Let's explore its key aspects:

3.1 Rta: The Cosmic Order

Rta is a fundamental concept in Vedic and later Hindu thought:

- It represents the principle of cosmic order, regularity, and harmony.

- Encompasses both natural and moral laws.

- Often associated with Truth (Satya) and Dharma (righteous living).

3.2 Rta in Vedic Literature

The concept of Rta is prominent in the Vedas:

- In the Rigveda, Rta is the principle that makes the sun shine, rivers flow, and seasons change.

- It's often personified as a cosmic power or associated with various deities.

- Rta is seen as the basis of cosmic and social harmony.

3.3 Rta and Karma

The concept of Rta is closely related to the law of Karma:

- Karma represents the principle of cause and effect in the moral realm.

- It can be seen as the moral dimension of Rta, ensuring cosmic justice.

3.4 Rta and Divine Will

In later Hindu thought, Rta is often associated with divine will:

- It's seen as the expression of the cosmic intelligence or Brahman.

- The gods are sometimes described as guardians or enforcers of Rta.

3.5 Rta and Human Life

The concept of Rta has implications for human conduct:

- Living in accordance with Rta is seen as the path to harmony and fulfilment.

- Rituals and ethical living are ways of aligning oneself with the cosmic order.

4. Comparative Analysis: Anthropic Principle and Cosmic Order

While the anthropic principle and the Hindu concept of cosmic order arise from very different contexts, they offer intriguing parallels and contrasts:

4.1 Purpose and Design

- The strong anthropic principle suggests a universe designed for life, resonating with the Hindu concept of a purposeful cosmic order.

- Both concepts grapple with the apparent orderliness and purposefulness of the cosmos.

4.2 The Role of Consciousness

- Some interpretations of the anthropic principle, particularly the participatory version, emphasize the role of conscious observers.

- Hindu thought often sees consciousness (Chit) as fundamental to the cosmos, intimately connected with Rta.

4.3 Moral Dimensions

- The Hindu concept of Rta encompasses moral laws, linking cosmic and ethical order.

- The anthropic principle is primarily concerned with physical laws, though it has sparked philosophical debates about purpose and ethics in the cosmos.

4.4 Scope and Applicability

- The anthropic principle is primarily concerned with explaining the physical conditions necessary for life.

- Rta is a broader concept, encompassing natural, social, and moral orders.

4.5 Epistemological Approaches

- The anthropic principle arises from scientific observation and reasoning.

- The concept of Rta comes from philosophical and spiritual insights, often claimed to be based on direct perception by sages.

5. Philosophical Implications

The study of the anthropic principle and cosmic order raises profound philosophical questions:

5.1 Teleology in the Cosmos

- Does the apparent fine-tuning of the universe suggest a purpose or goal?

- How do we reconcile purposefulness with scientific materialism?

5.2 The Nature of Physical Laws

- Are the laws of physics necessary, contingent, or chosen?

- How does the concept of multiple universes affect our understanding of physical laws?

5.3 The Place of Life and Consciousness in the Cosmos

- Is the emergence of life and consciousness a cosmic imperative or a happy accident?

- How central is consciousness to the nature of reality?

5.4 Divine Action and Natural Law

- How might concepts of divine action or cosmic intelligence relate to our understanding of natural laws?

- Can the regularity of natural laws be reconciled with the idea of divine intervention?

5.5 Anthropocentrism and Cosmic Humility

- Does the anthropic principle or the concept of cosmic order unduly prioritize human existence?

- How do we balance the seeming significance of consciousness with cosmic humility?

6. Scientific and Spiritual Perspectives on Cosmic Purpose

Both scientific and spiritual traditions have grappled with questions of cosmic purpose:

6.1 Scientific Views on Purpose

- Emergent Purpose: The idea that purpose can emerge from purposeless physical processes.

- Apparent Design: Explaining apparent design through natural selection and other non-teleological mechanisms.

6.2 Hindu Perspectives on Cosmic Purpose

- Lila: The concept of the cosmos as divine play.

- Spiritual Evolution: The idea of the cosmos evolving towards higher states of consciousness.

6.3 Integrating Perspectives

- Complementary Descriptions: Viewing scientific and spiritual perspectives as complementary rather than contradictory.

- Panpsychism and Neutral Monism: Philosophical approaches that might bridge scientific and spiritual views.

Panpsychism and neutral monism are two philosophical theories that are often considered incompatible.[36]

7. Ethical and Existential Implications

Our understanding of cosmic order and purpose has profound implications for how we live our lives:

7.1 Environmental Ethics

---

[36] *https://academic.oup.com/book/11114/*

- How does our view of cosmic order inform our relationship with nature?

- The concept of the Earth as a finely-tuned system (Gaia hypothesis) and its ethical implications.

7.2 Social and Political Philosophy

- How might concepts of cosmic order inform our ideas about social organization?

- The dangers of misusing ideas of cosmic order to justify social hierarchies or inequalities.

7.3 Personal Meaning and Purpose

- How do different views of cosmic order affect our sense of personal meaning and purpose?

- Navigating between cosmic significance and cosmic insignificance.

7.4 Attitude Towards Scientific Inquiry

- How do concepts of cosmic order affect our approach to scientific research?

- Balancing the search for underlying order with openness to new discoveries.

8. The Anthropic Principle, Cosmic Order, and the Future of Cosmology

As we look to the future of cosmological research, the anthropic principle and concepts of cosmic order continue to play important roles:

8.1 Multiverse Theories

- The role of the anthropic principle in evaluating and interpreting multiverse theories.

- Philosophical and methodological challenges in studying multiverses.

8.2 The Search for a Theory of Everything

- How concepts of cosmic order might inform the search for a unified theory of physics.

- The place of consciousness in a potential Theory of Everything.

8.3 Astrobiology and the Search for Extraterrestrial Life

- How the anthropic principle informs our search for life elsewhere in the universe.

- Implications of finding (or not finding) life elsewhere for our understanding of cosmic purpose.

8.4 Artificial Intelligence and the Nature of Mind

- How the development of AI might inform our understanding of consciousness and its place in the cosmos.

- Potential implications for the anthropic principle and concepts of cosmic order.

9. Conclusion: Towards a Holistic Understanding of Cosmic Order

As we've explored in this chapter, both the anthropic principle in modern cosmology and the concept of Rta in Hindu philosophy offer profound insights into the nature of cosmic order and purpose. While arising from very different contexts and methodologies, both perspectives grapple with the apparent fine-tuning and purposefulness we observe in the cosmos.

The anthropic principle challenges us to consider why the universe appears so exquisitely calibrated for the emergence of life and consciousness. It raises deep questions about the nature of physical laws, the possibility of multiple universes, and the role of observers in the cosmos. The Hindu concept of Rta, on the other hand, presents a vision of inherent cosmic order encompassing both natural and moral laws, suggesting an underlying harmony and purpose in the universe.

While these perspectives may seem disparate, they both encourage us to think deeply about our place in the cosmos and the nature of reality itself. The apparent fine-tuning of the universe, whether viewed through the lens of the anthropic principle or Rta, invites us to consider questions of purpose, design, and the fundamental nature of existence.

As we continue to probe the mysteries of the universe, both through scientific inquiry and contemplative insight, we may find new ways to synthesize these perspectives. Perhaps a more holistic understanding of cosmic order will emerge, one that honors both the rigorous observations of science and the profound insights of spiritual traditions.

In the next chapter, we will explore another fundamental aspect of our quest to understand cosmic origins: the search for unification in both scientific theories and spiritual philosophies.

9. Unification Theories – Science's Search for the Ultimate

1. Introduction: The Quest for Unity

Throughout the history of human thought, there has been a persistent drive to uncover the fundamental unity underlying the apparent diversity of the world. This quest for unification is evident in both scientific endeavors and spiritual philosophies. In modern physics, this search has led to increasingly comprehensive theories that attempt to explain diverse phenomena under a single framework. In Hindu philosophy, particularly in Advaita Vedanta, the concept of non-dualism (Advaita) represents the pinnacle of this unifying vision, positing an ultimate reality beyond all distinctions.

In this chapter, we will explore the parallel pursuits of unity in science and Hindu philosophy. We'll examine the development of unification theories in physics, from the unification of electricity and magnetism to the ongoing search for a Theory of Everything. We'll also delve into the concept of Advaita or non-dualism in Hindu thought, exploring its philosophical underpinnings and spiritual implications.

By bringing these two approaches into dialogue, we aim to uncover intriguing parallels and potential areas of convergence between scientific and spiritual quests for ultimate reality. This exploration may offer new perspectives on the nature of the cosmos and our place within it, challenging us to think deeply about the fundamental unity that may underlie all of existence.

2. Unification Theories in Modern Physics

The history of physics is marked by successive unifications, each bringing together seemingly disparate phenomena under more comprehensive frameworks. Let's explore this journey towards unification:

2.1 Early Unifications

2.1.1 Newton's Universal Gravitation

- Unified celestial and terrestrial mechanics

- Showed that the same laws govern the fall of an apple and the orbits of planets

2.1.2 Maxwell's Electromagnetism

- Unified electricity and magnetism

- Showed that light is an electromagnetic wave

2.2 Relativity: Unifying Space and Time

2.2.1 Special Relativity

- Unified space and time into spacetime

- Showed the equivalence of mass and energy ($E=mc^2$)

2.2.2 General Relativity

- Unified gravity with spacetime geometry

- Described gravity as the curvature of spacetime

2.3 Quantum Mechanics: Unifying the Microscopic World

- Provided a unified description of matter and energy at the subatomic level

- Introduced wave-particle duality, unifying particle and wave descriptions of matter and light

2.4 The Standard Model: Unifying Particles and Forces

- Describes three of the four fundamental forces (strong, weak, and electromagnetic) and all known elementary particles

- Unifies the electromagnetic and weak forces into the electroweak force

2.5 Beyond the Standard Model: The Search for Greater Unification

2.5.1 Grand Unified Theories (GUTs)

- Attempt to unify the strong, weak, and electromagnetic forces

- Predict phenomena like proton decay (not yet observed)

Proton decay is a phenomenon that is predicted by several grand unified theories (GUTs). GUTs are theories that unify the strong, weak, and electromagnetic forces of nature.[37]

2.5.2 Supersymmetry

- Proposes a symmetry between fermions and bosons

- Could potentially unify matter particles with force-carrying particles

2.5.3 String Theory and M-Theory

- Attempt to unify all fundamental forces, including gravity

- Propose that all particles are vibrations of tiny strings or membranes in higher-dimensional space

2.5.4 Loop Quantum Gravity

- Attempts to reconcile general relativity with quantum mechanics[38]

---

[37] *https://en.wikipedia.org/wiki/Proton_decay*
[38] *https://en.wikipedia.org/wiki/Loop_quantum_gravity*

- Describes spacetime as a network of loops at the Planck scale

2.6 The Ultimate Unification: Theory of Everything

- The holy grail of physics: a single theory explaining all physical aspects of the universe

- Challenges include reconciling quantum mechanics with general relativity and explaining phenomena like dark matter and dark energy

3. Non-Dualism (Advaita) in Hindu Philosophy

Advaita Vedanta, one of the most influential schools of Hindu philosophy, presents a radical vision of non-dualism. Let's explore its key concepts and development:

3.1 The Concept of Advaita

- Advaita means "not two," asserting the fundamental oneness of all reality

- Proposes that Brahman (ultimate reality) alone is real, and the perceived multiplicity is ultimately an illusion (maya)

3.2 Historical Development of Advaita

3.2.1 Upanishadic Roots

- Early expressions of non-dualism in texts like the Chandogya and Brihadaranyaka Upanishads

- The Mahavakyas (great sayings) like "Tat Tvam Asi" (That Thou Art) expressing the unity of individual self and ultimate reality

3.2.2 Gaudapada and the Mandukya Karika

- Systematic exposition of Advaita ideas

- Introduction of the concept of Ajativada (non-origination)

3.2.3 Adi Shankara

- Considered the most prominent exponent of Advaita Vedanta

- Developed a comprehensive philosophical system based on the Upanishads, Bhagavad Gita, and Brahma Sutras

3.3 Key Concepts in Advaita Vedanta

3.3.1 Brahman

- The one ultimate reality, beyond all attributes and distinctions

- Often described as Sat-Chit-Ananda (Existence-Consciousness-Bliss)

3.3.2 Atman

- The individual self, ultimately identical with Brahman

- The realization of this identity is considered liberation (moksha)

3.3.3 Maya

- The principle of illusion that creates the appearance of multiplicity

- Not separate from Brahman but Brahman's power of self-limitation

3.3.4 Levels of Reality

- Paramarthika Satya: Absolute reality (Brahman)

- Vyavaharika Satya: Empirical reality (the world of everyday experience)

- Pratibhasika Satya: Illusory reality (like dreams or misperceptions)

3.3.5 Jiva

- The individual soul, experiencing itself as separate due to ignorance (avidya)

- In reality, not different from Brahman

3.4 The Path to Realization in Advaita

- Shravana: Listening to the teachings

- Manana: Reflection and reasoning

- Nididhyasana: Deep meditation and contemplation

- The goal: Direct realization of one's true nature as Brahman

4. Comparative Analysis: Unification in Physics and Non-Dualism

While scientific unification theories and Advaita philosophy arise from very different contexts, they offer intriguing parallels:

4.1 The Search for Fundamental Unity

- Both approaches seek to understand reality in terms of an underlying unity

- Physics progresses towards ever more comprehensive theories; Advaita posits an ultimate, all-encompassing reality

4.2 Levels of Description

- Physics recognizes different levels of description (e.g., classical, quantum)

- Advaita proposes different levels of reality (Paramarthika, Vyavaharika, Pratibhasika)

4.3 The Role of the Observer

- Quantum mechanics emphasizes the role of the observer in measurement

- Advaita sees the ultimate observer (witness consciousness) as identical with Brahman

4.4 Transcending Ordinary Perception

- Both approaches suggest that ultimate reality transcends ordinary sensory perception

- Advanced physics theories often defy everyday intuition; Advaita posits a reality beyond the grasp of the ordinary mind

4.5 The Nature of Time and Space

- Modern physics describes spacetime as a unified entity

- Advaita sees time and space as part of the illusory manifestation (maya)

5. Philosophical Implications

The pursuit of unification in science and philosophy raises profound questions:

5.1 The Limits of Knowledge

- Can human minds comprehend ultimate reality?

- Are there fundamental limits to scientific unification or philosophical understanding?

5.2 The Nature of Consciousness

- How does consciousness fit into unified scientific theories?

- Is consciousness fundamental, as suggested by Advaita?

5.3 Reductionism vs. Holism

- Does unification necessarily imply reductionism?

- How can we balance reductionist and holistic approaches to understanding reality?

5.4 The Role of Mathematics

- Is mathematics discovered or invented?

- The concept of the universe as fundamentally mathematical and its parallels with Advaitic thought

5.5 Free Will and Determinism

- How do unified theories impact our understanding of causality and free will?

- Reconciling apparent individual agency with the concept of a unified reality

6. Challenges and Critiques

Both unification theories in physics and non-dual philosophies face significant challenges:

6.1 Challenges to Physical Unification Theories

- Lack of experimental evidence for some advanced theories (e.g., string theory)

- Difficulty in reconciling quantum mechanics and general relativity

- The problem of infinities in quantum field theories

6.2 Critiques of Advaita Vedanta

- The problem of accounting for the apparent reality of the world

- Ethical implications of a philosophy that sees the world as ultimately illusory

- Challenges from other schools of Indian philosophy (e.g., Dvaita, Vishishtadvaita)

6.3 General Challenges to Unification

- The risk of oversimplification or forced unification

- The possibility that reality might be fundamentally pluralistic

- The challenge of verifying or falsifying ultimate unification claims

7. Practical and Existential Implications

How do these unifying visions affect our understanding of life and our place in the cosmos?

7.1 Ethical Considerations

- How might a unified view of reality inform our ethical frameworks?

- The concept of interconnectedness and its ethical implications

7.2 Environmental Perspectives

- How do unification theories and non-dualism inform our relationship with nature?

- The Gaia hypothesis and its resonance with unifying philosophies

The Gaia hypothesis, also known as the Gaia theory, Gaia paradigm, or Gaia principle, is a model that proposes that living organisms and their nonliving surroundings on Earth form a self-regulating system that maintains conditions for life.[39]

7.3 Personal Identity and the Self

- How do these perspectives challenge our ordinary sense of self?

- Navigating between individual identity and universal unity

7.4 Approach to Life and Death

- How might a unified vision of reality affect our understanding of life and death?

- The concept of immortality in light of unification theories and non-dualism

---

[39] *https://courses.seas.harvard.edu/climate/eli/Courses/EPS281r/Sources/Gaia/Gaia-hypothesis-wikipedia.pdf*

8. Bridging Science and Spirituality

The parallel quests for unification in science and spirituality offer potential bridges between these domains:

8.1 Complementary Approaches

- Viewing scientific and spiritual approaches as complementary ways of investigating reality

- The potential for mutual enrichment between scientific and contemplative methods

8.2 Consciousness Studies

- The emerging field of consciousness studies as a meeting ground for scientific and spiritual inquiries

- Potential for integrating first-person (contemplative) and third-person (scientific) approaches

8.3 Holistic Worldviews

- The development of more integrated worldviews that honor both scientific and spiritual insights

- Examples of thinkers attempting to bridge science and spirituality (e.g., David Bohm, Ken Wilber)

8.4 Methodological Cross-Pollination

- What scientific methods might benefit spiritual inquiry?

- What contemplative insights might inform scientific research?

9. The Future of Unification

As we look to the future, several trends and possibilities emerge:

9.1 Advances in Fundamental Physics

- Potential breakthroughs in quantum gravity or other unifying theories

- The role of advanced technology (e.g., particle accelerators, gravitational wave detectors) in testing unification theories

9.2 Computational Approaches

- The increasing role of computational methods in advancing unification theories

- Simulations of unified theories and their implications

9.3 Interdisciplinary Integration

- Growing integration of physics with other fields like biology, neuroscience, and information theory

- Potential for more comprehensive theories that unify not just physical forces but broader aspects of reality

9.4 Evolving Philosophical Frameworks

- Development of new philosophical frameworks that can accommodate both scientific and contemplative insights

- The potential for a "philosophy of everything" that complements a "theory of everything"

10. Conclusion: Towards a More Unified Understanding of Reality

As we've explored in this chapter, both modern physics and Advaita Vedanta represent profound quests for ultimate unity. While they approach this search from vastly different perspectives and methodologies, both challenge us to transcend our ordinary perceptions and conceptions of reality.

The journey of physics towards ever more comprehensive unification theories reveals a cosmos of astounding interconnectedness and underlying simplicity. From Maxwell's unification of electricity and magnetism to the ongoing search for a Theory of Everything, this quest has continually expanded our understanding of the universe.

Similarly, the non-dual vision of Advaita Vedanta presents a radical understanding of reality as fundamentally one. By positing Brahman as the sole reality and the apparent multiplicity as a product of maya, it offers a philosophical framework for understanding ultimate unity.

While these approaches differ significantly in their methods and specific claims, they both point towards a reality that is more unified, more interconnected, and more profound than our everyday experience suggests. Both challenge us to expand our understanding, to question our assumptions, and to consider the possibility of a deeper, more fundamental unity underlying all of existence.

As we continue our exploration of cosmic origins, bridging the Big Bang and Brahman, this quest for unification serves as a powerful reminder of the depth and mystery of the cosmos. It encourages us to remain open to new discoveries, to push the boundaries of our understanding, and to seek a more comprehensive and integrated vision of reality.

In the next chapter, we will explore the limits of knowledge in both scientific and spiritual inquiries, examining how these limitations shape our understanding of cosmic origins and ultimate reality.

10. The Limits of Knowledge

Introduction: The Boundaries of Understanding

As we approach the culmination of our exploration into cosmic origins, bridging the scientific perspective of the Big Bang with the Hindu concept of Brahman, we find ourselves confronting a profound and humbling reality: the limits of human knowledge. Both science and spirituality, in their quests to understand the nature of reality, eventually encounter boundaries beyond which our current understanding falters.

In science, these limits manifest in various ways: the uncertainty principle in quantum mechanics, the observer effect, the limitations of our measuring devices, and the sheer complexity of the cosmos. In Hindu philosophy, particularly in concepts like Neti Neti ("not this, not that") and the ineffability of Brahman, we find a recognition of the ultimate limitations of conceptual thought in grasping ultimate reality.

This chapter will explore these limits of knowledge from both scientific and spiritual perspectives. We will examine how these limitations shape our understanding of cosmic origins and the nature of reality. By bringing these two approaches into dialogue, we aim to develop a more nuanced appreciation of what can be known, what might be unknowable, and how we might navigate the vast terrain of the unknown.

Our exploration of these limits is not meant to discourage the pursuit of knowledge, but rather to instil a sense of humility and wonder in the face of the cosmos's profound mysteries. It may also point us towards new ways of knowing that transcend the limitations of our current paradigms.

2. Scientific Limits of Knowledge

Modern science, despite its remarkable successes, encounters various fundamental limits to knowledge. Let's explore some of these limitations:

2.1 Quantum Uncertainty

2.1.1 The Heisenberg Uncertainty Principle

- Fundamental limit on the precision with which certain pairs of physical properties can be determined

- Implications for our ability to predict the behavior of quantum systems

2.1.2 Wave Function Collapse and Measurement

- The measurement problem in quantum mechanics

- Various interpretations and their implications for the nature of reality and our knowledge of it

2.2 Limits in Cosmology

2.2.1 The Cosmic Horizon

- The limit of the observable universe due to the finite speed of light and the expansion of space

- Implications for our ability to test cosmological theories

2.2.2 The Planck Era

- The earliest period of the universe, where our current physical theories break down

- Challenges in understanding the initial conditions of the universe

2.2.3 Dark Matter and Dark Energy

- Limitations in our understanding of the majority of the universe's matter and energy content

2.3 Chaotic Systems and Predictability

- The butterfly effect and the limits of long-term prediction in complex systems

- Implications for weather forecasting, climate modeling, and understanding biological systems

2.4 Gödel's Incompleteness Theorems

- Limitations in formal logical systems

- Implications for mathematics and our ability to construct a complete and consistent theory of everything

2.5 The Problem of Induction

- Limitations of inferring general rules from specific observations

- The challenge of justifying scientific laws based on finite observations

2.6 Technological and Practical Limits

- Limitations of measuring devices and experimental setups

- Practical constraints on energy and scale in particle physics experiments

2.7 Cognitive and Perceptual Limits

- Limitations of human sensory systems and cognitive capacities

- The challenge of conceptualizing realities beyond our everyday experience (e.g., higher dimensions, quantum phenomena)

3. Limits of Knowledge in Hindu Philosophy

Hindu philosophy, particularly in its more mystical and non-dualistic forms, recognizes profound limits to conceptual knowledge. Let's explore some key ideas:

3.1 Neti Neti (Not This, Not That)

- A method of negation used to approach the understanding of Brahman

- Recognition that ultimate reality transcends all descriptions and concepts

3.2 Maya and the Limits of Perception

- The concept of Maya as the power that veils ultimate reality

- Limitations of sensory and mental perception in grasping the true nature of reality

3.3 Ineffability of Brahman

- The ultimate reality (Brahman) as beyond language and thought

- The limitations of conceptual knowledge in realizing non-dual awareness

3.4 Avidya (Ignorance)

- The fundamental ignorance that prevents direct realization of ultimate reality

- Distinction between intellectual knowledge and experiential realization

3.5 Limits of Scriptural Knowledge

- Recognition in many Hindu traditions that even scriptural knowledge is ultimately limited

- The need for direct experience (anubhava) beyond textual learning

3.6 The Four States of Consciousness

- Waking, dreaming, deep sleep, and turiya (the fourth state)

- Turiya as a state beyond the grasp of ordinary consciousness and knowledge

3.7 Yogic Epistemology

- Different modes of knowing: pratyaksha (perception), anumana (inference), shabda (testimony)

- Recognition of higher modes of cognition through yogic practices

4. Comparative Analysis: Scientific and Hindu Approaches to the Limits of Knowledge

While arising from very different contexts, scientific and Hindu approaches to the limits of knowledge offer some intriguing parallels:

4.1 The Role of the Observer

- **Quantum mechanics**: The observer effect and its implications for objective knowledge

- **Advaita Vedanta**: The ultimate observer (witness consciousness) as identical with the observed reality

4.2 Beyond Ordinary Perception

- **Science:** Realities beyond human sensory perception (e.g., quantum realm, cosmic scales)

- **Hindu thought**: Realities beyond ordinary states of consciousness (e.g., turiya)

4.3 Limits of Language and Concept

- **Science**: Challenges in describing quantum phenomena or higher dimensions in everyday language

- **Hindu philosophy**: The ineffability of Brahman and limitations of conceptual thought

4.4 Knowing vs. Being

- **Science**: The distinction between theoretical knowledge and experimental reality

- **Hindu thought**: The distinction between intellectual understanding and experiential realization

4.5 Fundamental Uncertainty

- **Quantum mechanics**: Inherent uncertainty at the quantum level

- **Hindu philosophy**: Fundamental unknowability of ultimate reality through ordinary means

5. Philosophical Implications

The recognition of the limits of knowledge raises profound philosophical questions:

5.1 The Nature of Reality

- Is reality fundamentally knowable, or are there aspects forever beyond our grasp?

- How do we navigate between pragmatic and ultimate views of reality?

5.2 The Role of Consciousness

- How does consciousness relate to the limits of knowledge?

- Can expanded states of consciousness transcend ordinary epistemic limits?

5.3 Truth and Knowledge

- What constitutes "truth" in light of these limitations?

- How do we evaluate competing knowledge claims given these limits?

5.4 The Purpose of Inquiry

- If ultimate knowledge is limited, what is the purpose of scientific or spiritual inquiry?

- How do we balance the pursuit of knowledge with the recognition of its limits?

5.5 Free Will and Determinism

- How do epistemic limits affect our understanding of causality and free will?

- Can we reconcile the appearance of choice with fundamental uncertainty or non-dual reality?

6. Navigating the Unknown: Strategies and Approaches

Given the limits of knowledge, how can we best approach the unknown? Both scientific and spiritual traditions offer strategies:

6.1 Scientific Approaches

6.1.1 Embracing Uncertainty

- Probabilistic thinking and statistical approaches

- Developing comfort with multiple interpretations and models

6.1.2 Pushing Technological Boundaries

- Developing more precise measuring tools and experimental techniques

- Leveraging computational power to explore complex systems

6.1.3 Interdisciplinary Collaboration

- Combining insights from multiple fields to tackle complex problems

- Developing new conceptual frameworks that transcend traditional disciplinary boundaries

6.1.4 Theoretical Speculation

- Developing theories that go beyond current experimental capabilities

- Using mathematical consistency and beauty as guides in theory development

6.2 Spiritual Approaches

6.2.1 Cultivating Direct Experience

- Meditation and contemplative practices to explore consciousness directly

- Emphasis on experiential knowledge beyond conceptual understanding

6.2.2 Developing Intuitive Wisdom

- Cultivating prajna or intuitive wisdom that transcends rational thought

- Practices to quiet the mind and access deeper levels of knowing

6.2.3 Embracing Mystery

- Cultivating a sense of wonder and humility in the face of the unknown

- Seeing limitations as invitations to deeper exploration rather than obstacles

6.2.4 Ethical and Compassionate Living

- Focus on right living and ethical behavior as a path to deeper understanding

- Cultivating compassion and interconnectedness to expand beyond individual limitations

7. Implications for Cosmology and Cosmic Origins

How do the limits of knowledge affect our understanding of cosmic origins?

7.1 The Limits of Cosmic Models

- Challenges in testing theories about the very early universe

- The problem of underdetermination in cosmological models

7.2 Multiverses and Cosmic Inflation

- Potential unknowability of realities beyond our cosmic horizon

- Philosophical implications of potentially unobservable cosmic domains

7.3 The Role of Consciousness in Cosmic Origins

- Debates about the anthropic principle and observer-dependent realities

- Parallels with Hindu concepts of cosmic consciousness and manifestation

7.4 Ultimate Origins

- The persistent question of "why is there something rather than nothing?"

- Limits in scientifically or philosophically resolving the question of ultimate origins

8. Ethical and Existential Implications

How does our recognition of the limits of knowledge affect how we live and act in the world?

8.1 Intellectual Humility

- Cultivating humility in the face of the vast unknown

- Balancing confidence in knowledge with openness to new discoveries

8.2 Dealing with Uncertainty

- Developing personal and societal strategies for navigating uncertainty

- The challenge of decision-making in the face of incomplete knowledge

8.3 The Value of Mystery

- Appreciating mystery as a source of wonder and continued exploration

- The role of the unknown in driving scientific and spiritual quests

8.4 Ethical Responsibility

- The ethical implications of acting on limited knowledge

- Balancing the pursuit of knowledge with ethical considerations

8.5 Meaning and Purpose

- Finding meaning in the face of fundamental uncertainty

- The role of limiting knowledge in shaping our sense of purpose and place in the cosmos

9. Bridging Science and Spirituality

The recognition of epistemic limits offers potential bridges between scientific and spiritual approaches:

9.1 Complementary Ways of Knowing

- Recognizing the value of both empirical and contemplative approaches

- Exploring how scientific and spiritual methods might complement each other

9.2 Cultivating Cognitive Flexibility

- Developing the ability to hold multiple perspectives simultaneously

- Appreciating both reductionist and holistic approaches to knowledge

9.3 Integrative Frameworks

- Developing philosophical frameworks that can accommodate both scientific and spiritual insights

- Exploring concepts like holomovement (David Bohm) or integral theory (Ken Wilber)

9.4 Dialogue and Cross-Pollination

- Fostering dialogue between scientists, philosophers, and contemplatives

- Exploring how insights from one domain might inform or inspire the other

10. The Future of Knowledge: Emerging Paradigms

As we look to the future, several trends and possibilities emerge in our approach to knowledge:

10.1 Quantum Cognition

- Applying quantum principles to understanding cognition and decision-making

- Potential for new models of mind that transcend classical limitations

10.2 Artificial Intelligence and Knowledge

- The role of AI in pushing the boundaries of human knowledge

- Philosophical implications of potentially superhuman AI

10.3 Expanded States of Consciousness

- Scientific study of altered states of consciousness and their epistemic potential

- Integration of contemplative insights into scientific frameworks

10.4 Participatory Epistemologies

- Developing frameworks that recognize the participatory nature of knowledge

- Exploring how the act of knowing shapes the known

10.5 Post-Materialist Science

- Emerging scientific paradigms that go beyond strict materialism

- Integrating consciousness and subjective experience into scientific models

11. Conclusion: Embracing the Mystery

As we conclude our exploration of the limits of knowledge, we find ourselves standing at the edge of a vast cosmic mystery. Both science and spirituality, in their deepest inquiries, lead us to this threshold where our current understanding gives way to the unknown.

The journey of science, from quantum uncertainty to cosmic horizons, reveals a universe far stranger and more mysterious than our everyday perception suggests. It shows us a reality where certainty gives way to probability, where the act of observation shapes the observed, and where our quest for knowledge continually unveils new mysteries.

Similarly, the spiritual insights of Hindu philosophy, particularly in concepts like Neti Neti and the ineffability of Brahman, point us towards a reality that transcends our conceptual frameworks. They suggest a depth to existence that cannot be fully captured by thought or language, inviting us into a direct experience of the mystery of being.

These limitations, far from being obstacles, can be seen as invitations. They invite us to cultivate intellectual humility, to remain open to new discoveries, and to approach the cosmos with a sense of wonder. They challenge us to develop new ways of knowing, to push the boundaries of our understanding, and to hold our knowledge lightly.

As we continue to probe the mysteries of cosmic origins, bridging the Big Bang and Brahman, the recognition of our epistemic limits serves as a powerful reminder of the vastness and depth of the cosmos. It encourages us to embrace both rigorous inquiry and open-minded wonder, to pursue knowledge passionately while remaining humble in the face of the ultimate mystery of existence.

In this space of "learned ignorance," where we know that we do not know, we may find a deeper wisdom. Here, scientific curiosity and spiritual awe can coexist and enrich each other, leading us towards a more comprehensive and nuanced understanding of our cosmic origins and our place in the universe.

156

11. Ethical Implications

1. Introduction: The Cosmic Context of Ethics

As we near the conclusion of our exploration into cosmic origins, bridging the scientific perspective of the Big Bang with the Hindu concept of Brahman, we find ourselves confronting a profound question: What are the ethical implications of our understanding of the cosmos? How does our view of the universe's origin and nature inform our moral frameworks and guide our actions?

This chapter will explore the ethical dimensions that emerge from our scientific and spiritual investigations into cosmic origins. We will examine how our understanding of the universe – its vastness, its age, its fundamental laws, and our place within it –

shapes our sense of meaning, purpose, and moral responsibility. Similarly, we will consider how Hindu concepts like Brahman, karma, and the interconnectedness of all beings inform ethical thinking and behavior.

By bringing these scientific and spiritual perspectives into dialogue, we aim to develop a more comprehensive and nuanced approach to ethics — one that is grounded in our best understanding of cosmic realities while also drawing on the profound insights of spiritual traditions. This cosmic ethics may offer new ways of addressing pressing global challenges, from environmental sustainability to social justice, and guide us towards a more harmonious relationship with each other and the universe at large.

Our exploration of cosmic ethics is not merely an academic exercise. In an age of global challenges and rapid technological advancement, the need for an expanded ethical framework that can encompass our cosmic context has never been more urgent. By considering the ethical implications of our cosmic understanding, we may find new inspiration and guidance for navigating the complexities of human existence on our small but precious planet.

2. Ethical Implications of Scientific Cosmology

Our scientific understanding of the cosmos has profound ethical implications. Let's explore some key areas:

2.1 Cosmic Perspective and Human Significance

2.1.1 The Pale Blue Dot

- Carl Sagan's reflection on the Earth as seen from deep space

- Implications for human humility and the preciousness of our planet

2.1.2 Cosmic Time Scales

- The brevity of human existence compared to cosmic time scales

- Implications for long-term thinking and intergenerational responsibility

2.2 The Anthropic Principle and Responsibility

- The fine-tuning of the universe for life

- Ethical implications of being possibly rare or unique observers in the cosmos

2.3 Interconnectedness in the Cosmos

- The shared origin of all matter in the Big Bang

- Ecological ethics based on the interconnectedness of all Earth systems

2.4 Evolution and Ethics

- Evolutionary origins of moral behavior

- Balancing evolutionary heritage with ethical aspirations

2.5 Entropy and Environmental Ethics

- The second law of thermodynamics and its implications for resource use

- Long-term sustainability in light of cosmic principles

2.6 Extraterrestrial Life and Expanding Moral Circles

- The possibility of life elsewhere in the universe

- Preparing ethical frameworks for potential contact with extraterrestrial intelligence

2.7 Quantum Entanglement and Moral Responsibility

- Implications of quantum interconnectedness for ethical thinking

- Challenging classical notions of individual autonomy

3. Ethical Dimensions in Hindu Cosmology

Hindu cosmology and philosophy offer rich resources for ethical reflection. Let's explore some key concepts:

3.1 Brahman and the Unity of Existence

- Ethical implications of the fundamental oneness of all reality

- Expanding circles of compassion based on non-dual awareness

3.2 Karma and Moral Responsibility

- The law of karma as a cosmic principle of moral causation

- Balancing individual responsibility with compassion

3.3 Dharma and Cosmic Order

- The concept of dharma as righteous living in harmony with cosmic order

- Ethical decision-making based on one's role and context

3.4 Ahimsa (Non-violence)

- The principle of non-harm extended to all beings

- Ecological and social applications of ahimsa

3.5 Yoga and Self-Transformation

- Ethical development through yogic practices

- The role of self-realization in ethical behavior

3.6 Cycles of Time and Long-term Thinking

- Yugas and cosmic cycles inspiring long-term ethical perspectives

- Balancing immediate concerns with cosmic time scales

3.7 Maya and Ethical Discernment

- Navigating ethical decisions in a world of appearances

- Balancing engagement with detachment

4. Comparative Analysis: Scientific and Hindu Approaches to Cosmic Ethics

While arising from different contexts, scientific and Hindu perspectives on cosmic ethics offer intriguing parallels and complementarities:

4.1 Interconnectedness and Unity

- Science: Ecological interconnectedness, shared cosmic origins

- Hinduism: Non-dual reality of Brahman, interconnectedness of all beings

4.2 Expanded Time Scales

- Science: Cosmic time scales, long-term consequences of actions

- Hinduism: Cycles of yugas, multiple lifetimes, karma across incarnations

4.3 Observer ship and Consciousness

- **Science:** Anthropic principle, the role of conscious observers

- **Hinduism**: Consciousness as fundamental, the ultimate observer (Atman/Brahman)

4.4 Ethical Evolution

- **Science**: Evolutionary origins of morality, potential for ethical progress

- **Hinduism**: Spiritual evolution, the refinement of karma across lifetimes

4.5 Universal Principles

- **Science**: Laws of physics as universal principles

- **Hinduism**: Dharma as cosmic order and ethical principle

5. Foundational Ethical Questions in Light of Cosmic Origins

Our understanding of cosmic origins raises fundamental ethical questions:

5.1 The Basis of Moral Value

- Does the vastness of the cosmos diminish or enhance the value of conscious experience?

- How does the potential rarity of life in the universe affect its moral status?

5.2 Free Will and Moral Responsibility

- How do deterministic laws of physics or the concept of karma affect our notion of free will?

- Reconciling moral responsibility with cosmic causality

5.3 The Nature of Good and Evil

- Are good and evil fundamental features of the cosmos or human constructs?

- The problem of evil in light of cosmic order or divine pla

5.4 Meaning and Purpose

- Can a purposeless universe as described by some scientific models provide a basis for meaning?

- How does the Hindu concept of lila (divine play) Inform our sense of cosmic purpose?

5.5 Justice and Cosmic Order

- How do we understand justice in light of the apparent indifference of the cosmos?

- The concept of cosmic justice in karma and its ethical implications

6. Applied Cosmic Ethics

How might our understanding of cosmic origins inform our approach to pressing ethical issues?

6.1 Environmental Ethics

- Stewardship of Earth as a rare life-bearing planet

- Long-term sustainability inspired by cosmic time scales

6.2 Social Justice

- Equality based on shared cosmic origins

- Expanding circles of moral concern inspired by cosmic unity

6.3 Bioethics

- Genetic engineering in light of cosmic evolution

- The moral status of potential artificial life or consciousness

6.4 Global Governance

- Cosmic perspective informing international cooperation

- Preparing for potential extraterrestrial contact

6.5 Technology Ethics

- Responsible innovation inspired by cosmic creativity

- Long-term consequences of technological development

6.6 Animal Rights

- Expanding moral consideration based on evolutionary kinship

- Ahimsa applied to human-animal relationships

6.7 Economic Systems

- Sustainable economics informed by cosmic principles

- Balancing growth with cosmic rhythms of creation and dissolution

7. Challenges and Critiques

The attempt to derive ethics from cosmic understanding faces several challenges:

7.1 The Is-Ought Problem

The is-ought problem is the idea that you can't derive an "ought" conclusion from "is" premises, or statements about what is. It was articulated by Scottish philosopher and historian David Hume.

Here's an example of the is-ought problem:

- Descriptive: Humans die if you electrocute them.
- Descriptive: Tim is human.
- Moral: Therefore, you ought not electrocute Tim.

This looks like a valid deductive argument, but the premises do not entail the conclusion.

The is-ought problem is central to the debate about empirical ethics. Most scientists agree that science can only describe the way things are, but never tell us how they ought to be.

The is-ought fallacy is when someone assumes that because things are a certain way, they should be that way. For example,

"We do not currently regulate the amount of nicotine in an individual cigarette; therefore we need not do this".

- The challenge of deriving normative statements from descriptive facts

- Navigating between cosmic "is" and ethical "ought"

7.2 Relativism vs. Universalism

- Relativism

The belief that there is no absolute truth, and that truth and morality are relative to the individual or culture. Relativists believe that people can have different views on what is moral and immoral. They may also believe that cultural practices should be respected, even if they are harmful to others.

- Universalism

The belief that some ideas have universal application, and that there is one fundamental truth. Universalists believe that moral truths do not vary from person to person or culture to culture. They may also believe that individuals should apply consistent moral principles regardless of cultural or situational variations.

Here are some examples of relativism and universalism:

- Ethical relativism

Some examples of ethical relativism include royals marrying family members, divorce from marriage as sinful, and widows remarrying as morally taboo.

- Universalism

An example of universalism is the United Nations' Universal Declaration of Human Rights, which asserts rights to all people regardless of culture or nationality.

- Balancing universal cosmic principles with cultural diversity

- The challenge of establishing universal ethics in a vast and diverse cosmos

7.3 Anthropocentrism

Anthropocentrism is the belief that humans are the most important beings on the planet and that other beings are only valuable if they serve humans. It can also be referred to as Humano centrism, human supremacy, or human exceptionalism.

- The risk of projecting human values onto the cosmos

- Balancing human concerns with broader cosmic perspectives

7.4 Fatalism and Disengagement

Fatalistic passiveness is a personality trait where a person accepts an inevitable outcome and becomes more passive in the face of events:

- Beliefs: Fatalists believe that the outcome is inevitable no matter what they do. They may say things like "There's nothing I could do about it" or "This was bound to happen".
- Feelings: Fatalism can lead to feelings of hopelessness, resignation, and unhappiness.
- Behavior: Fatalists may shrink their realm of choice and become more passive.

Fatalism can manifest in different ways depending on the situation and the person's culture:

- Collectivist cultures

In cultures with poor economic development, fatalism is a cognitive schema that involves accepting an irremediable destiny.

- Individualist cultures

In highly developed cultures, fatalism can manifest as a mood of uncertainty, insecurity, and helplessness.

- Attitudes

People may have a fatalistic attitude towards their job, believing they can't change their situation.

- Valences

Fatalism can be neutral or pessimistic. Neutral fatalism is the belief that you can't influence the outcome, regardless of whether it's good, bad, or indifferent. Pessimistic fatalism is the expectation that nothing good will happen.

- Addressing the risk of ethical passivity in face of cosmic vastness

- Maintaining ethical motivation in light of cosmic time scales

7.5 Competing Interpretations

- Navigating different interpretations of cosmic data or spiritual insights

- Resolving conflicts between scientific and spiritual ethical imperatives

8. Towards a Cosmic Ethic

Drawing on both scientific and Hindu insights, we can outline some principles for a cosmic ethic:

8.1 Cosmic Humility

- Cultivating humility in face of the universe's vastness and complexity

- Balancing human agency with cosmic perspective

8.2 Expanded Compassion

- Extending empathy and concern to all sentient beings

- Recognizing the shared cosmic heritage of all life

8.3 Long-term Stewardship

- Taking responsibility for the long-term consequences of our actions

- Thinking in terms of cosmic time scales

8.4 Harmony with Cosmic Rhythms

- Aligning human activities with natural and cosmic cycles

- Balancing progress with sustainability

8.5 Consciousness Cultivation

- Recognizing the central role of consciousness in the cosmos

- Ethical development through expansion of awareness

8.6 Creative Participation

- Viewing ethical action as participation in cosmic creativity

- Balancing acceptance of cosmic laws with creative engagement

8.7 Unity in Diversity

- Celebrating the diversity of life and culture as expressions of cosmic creativity

- Recognizing underlying unity amidst apparent diversity

9. Practical Applications of Cosmic Ethics

How might cosmic ethics be applied in everyday life and societal structures?

9.1 Education

- Integrating cosmic perspective into educational curricula

- Fostering wonder and ethical responsibility through cosmic education

9.2 Policy Making

- Incorporating long-term cosmic thinking into policy decisions

- Balancing immediate needs with long-term cosmic sustainability

9.3 Personal Development

- Practices for cultivating cosmic awareness and ethical behavior

- Aligning personal goals with cosmic perspectives

9.4 Business and Economics

- Developing business models aligned with cosmic ethical principles

- Redefining success and growth in cosmic terms

9.5 Arts and Culture

- Expressing cosmic ethical insights through various art forms

- Cultivating a culture of cosmic citizenship

9.6 Science and Technology

- Ethical guidelines for scientific research based on cosmic principles

- Developing technologies in harmony with cosmic ethics

9.7 Spiritual Practices

- Integrating scientific insights into spiritual and contemplative practices

- Developing practices that foster cosmic ethical awareness

10. The Future of Cosmic Ethics

As we look to the future, several trends and possibilities emerge:

10.1 Astro ethics

- Developing ethical frameworks for space exploration and utilization

- Preparing for potential contact with extraterrestrial intelligence

10.2 Evolutionary Ethics

- Consciously guiding human ethical evolution

- Integrating cosmic evolutionary principles into ethical development

10.3 Quantum Ethics

- Exploring ethical implications of quantum phenomena like entanglement

- Developing ethical frameworks that account for quantum reality

10.4 Cosmic Citizenship

- Fostering a sense of belonging to the cosmos

- Developing institutions and practices of cosmic citizenship

10.5 Integrative Ethical Frameworks

- Synthesizing scientific, philosophical, and spiritual ethical insights

- Developing comprehensive ethical systems for a cosmic age

11. Conclusion: Our Cosmic Ethical Imperative

As we conclude our exploration of the ethical implications of cosmic origins, we find ourselves standing at a unique juncture in human history. Our scientific understanding of the cosmos has revealed a universe of staggering vastness, complexity, and interconnectedness. At the same time, spiritual traditions like

Hinduism offer profound insights into the nature of reality and our place within it.

The convergence of these scientific and spiritual perspectives presents us with a cosmic ethical imperative. We are called to expand our moral horizons to encompass the full scope of our cosmic context. This expanded ethics challenges us to think in vast time scales, to recognize our fundamental interconnectedness with all of existence, and to take responsibility for our role in the cosmic drama.

From the scientific perspective, we are the universe becoming aware of itself, the means by which the cosmos can contemplate its own existence. This unique position brings with it a profound responsibility – to understand, to preserve, and to contribute positively to the ongoing evolution of cosmic complexity and consciousness.

From the Hindu perspective, we are expressions of the divine cosmic play, with the potential to realize our fundamental unity with Brahman. This realization brings with it the ethical imperative to act from a place of cosmic awareness, to see the divine in all beings, and to align our actions with the cosmic dharma.

Bridging these perspectives, we can envision an ethic that honors both the objective insights of science and the subjective depths of spiritual wisdom. This cosmic ethic calls us to act with humility in the face of cosmic vastness, with wonder at the miracle of existence, with compassion for all sentient beings, and with responsibility for the long-term flourishing of life and consciousness.

As we face global challenges that threaten the future of our species and our planet, this cosmic ethic offers a expanded perspective that may guide us towards more sustainable, compassionate, and wise courses of action. It reminds us that our brief human story is part of a grand cosmic narrative, and that our choices and actions ripple out into the vast ocean of cosmic time and space.

In embracing this cosmic ethical perspective, we may find not only a guide for navigating the complexities of human existence but also a source of profound meaning and connection. We are, after all, children of the cosmos – whether understood through the lens of the Big Bang or Brahman – and our ethical journey is nothing less than the universe exploring its own moral dimensions through us.

As we move forward, may we carry with us this sense of cosmic ethical responsibility, allowing it to inform our choices, inspire our actions, and guide us towards a future worthy of our cosmic heritage.

12. Synthesis – Toward a Holistic Cosmic Vision

Introduction: Bridging Paradigms

As we reach the culmination of our exploration into cosmic origins, bridging the scientific perspective of the Big Bang with the Hindu concept of Brahman, we stand at a unique vantage point. Throughout this journey, we have delved into the depths of modern cosmology and the profound insights of Hindu philosophy, examining their parallels, contrasts, and potential complementarities. Now, we face the challenge and opportunity of synthesis – the creation of a more holistic cosmic vision that honors both the rigorous empiricism of science and the deep intuitions of spiritual wisdom.

This chapter aims to weave together the threads we have explored, seeking a unified understanding that transcends the boundaries of individual disciplines. We will examine how the insights from modern physics, cosmology, Hindu philosophy, and other perspectives we have encountered might be integrated into a more comprehensive worldview. This synthesis is not about forcing disparate ideas into an artificial unity, but rather about finding genuine points of convergence and complementarity, while also acknowledging and respecting differences.

Our goal is to develop a cosmic vision that is intellectually coherent, emotionally satisfying, and practically relevant. This holistic perspective should be capable of addressing the big questions of existence – the nature of reality, the origin and destiny of the cosmos, the place of consciousness in the universe, and the meaning and purpose of life – in a way that integrates the best of our scientific and spiritual understanding.

As we embark on this synthesis, we recognize that we are participating in a grand dialogue that has spanned millennia and cultures. Our effort is not to provide final answers, but to contribute to an ongoing conversation about the nature of reality and our place within it. In doing so, we hope to open new pathways of understanding and inspire further exploration of the profound mystery of cosmic origins and the nature of existence itself.

2. Recap: Key Insights from Science and Hindu Philosophy

Before we attempt a synthesis, let's briefly recap some of the key insights we've explored from both scientific and Hindu perspectives:

2.1 Scientific Insights

2.1.1 The Big Bang and Cosmic Evolution

- The universe had a beginning about 13.8 billion years ago

- Cosmic evolution from simple to complex structures

2.1.2 Quantum Mechanics

- Fundamental uncertainty and interconnectedness at the quantum level

- The role of the observer in quantum phenomena

2.1.3 Relativity

- Space and time as interconnected aspects of spacetime

- The malleable nature of spacetime in relation to matter and energy

2.1.4 Emergence and Complexity

- The emergence of complex systems from simple rules

- Self-organization and the evolution of cosmic structures

2.1.5 Anthropic Principle

- The apparent fine-tuning of the universe for the emergence of life

2.2 Insights from Hindu Philosophy

2.2.1 Brahman

- The concept of an ultimate, unchanging reality underlying all existence

2.2.2 Maya

- The world of appearances as a manifestation of the divine

2.2.3 Atman

- The individual self as fundamentally identical with Brahman

2.2.4 Karma and Reincarnation

- The law of cause and effect operating across multiple lifetimes

2.2.5 Yoga and Consciousness

- Techniques for exploring and expanding consciousness

2.2.6 Cyclical Time

- The concept of cosmic cycles (yugas) and recurring creation and dissolution

3. Points of Convergence

Despite their different methodologies and contexts, science and Hindu philosophy show several intriguing points of convergence:

3.1 Unity and Interconnectedness

- **Science**: Quantum entanglement, ecological interdependence

- **Hinduism**: Non-dual nature of Brahman, interconnectedness of all beings

3.2 The Role of the Observer

- **Science**: The observer effect in quantum mechanics

- **Hinduism**: The ultimate observer (Atman/Brahman) as the ground of all experience

3.3 Levels of Reality

- **Science**: Different levels of description (quantum, classical, emergent phenomena)

- **Hinduism**: Levels of reality (Paramarthika, Vyavaharika, Pratibhasika)

3.4 The Limits of Knowledge

- **Science**: Uncertainty principle, Gödel's incompleteness theorems

- **Hinduism**: The ultimate ineffability of Brahman, limitations of conceptual knowledge

3.5 Vast Time Scales

- **Science**: Billions of years of cosmic evolution

- **Hinduism**: Vast cycles of cosmic time (yugas, kalpas)

3.6 The Emergence of Complexity

- **Science**: The evolution of complex structures and life

- **Hinduism**: The manifestation of diverse forms from the formless Brahman

4. Toward a Unified Cosmic Vision

Building on these points of convergence, we can outline elements of a more holistic cosmic vision:

4.1 Fundamental Unity

A core principle of our unified vision is the fundamental unity underlying all existence. This aligns with both the scientific quest for unified theories and the Hindu concept of Brahman. We might conceptualize the universe as a single, interconnected whole, manifesting in diverse forms but fundamentally one.

4.2 Dynamic Manifestation

Within this unity, we recognize a dynamic process of manifestation. The scientific narrative of cosmic evolution, from the Big Bang through the formation of galaxies, stars, and life, can be seen as paralleling the Hindu concept of divine manifestation or lila (divine play). This process is characterized by:

- Emergence of complexity from simplicity

- Self-organization and evolution

- Cycles of creation, preservation, and dissolution

4.3 Consciousness as Fundamental

Drawing on both quantum theories of consciousness and the Hindu concept of Chit (consciousness as an aspect of Brahman), our unified vision posits consciousness as a fundamental aspect of reality, not merely an emergent property of complex brains.

This perspective sees awareness as intrinsic to the cosmos, manifesting in various degrees throughout nature.

4.4 Multi-layered Reality

Our holistic vision acknowledges multiple levels or layers of reality, integrating scientific descriptions of different scales (quantum, classical, cosmic) with Hindu concepts of levels of reality. This multi-layered model allows for different, complementary descriptions of reality, each valid in its own domain.

4.5 Participatory Universe

Inspired by both the role of the observer in quantum mechanics and the Hindu concept of lila, we envision a participatory universe where consciousness plays an active role in the unfolding of reality. This perspective sees humans (and potentially other conscious beings) as active participants in cosmic evolution, not merely passive observers.

In chapter 3 of Bhagvad Gita, Lord Krishna talks about the devatas (Celestial beings) who manage the outer world. Constant interaction with these devatas is the responsibility of a human being though yagnas (sacrificial rituals). Thus both the devatas and the humans manage the show of the world.

4.6 Ethical Dimension

Our unified vision incorporates an ethical dimension, recognizing moral development as an integral part of cosmic evolution. This aligns with both evolutionary ethics and Hindu concepts of dharma and karma, seeing ethical behavior as alignment with cosmic principles.

4.7 Transcendent Yet Immanent Mystery

While striving for understanding, our holistic vision maintains a sense of mystery and transcendence. It recognizes that the deepest nature of reality may always elude complete conceptual grasp, echoing both the limits of scientific knowledge and the Hindu notion of the ultimate ineffability of Brahman.

5. Implications of a Unified Cosmic Vision

This holistic cosmic vision has profound implications for various aspects of human thought and life:

5.1 Epistemology

- Integration of empirical, rational, and contemplative modes of knowing

- Recognition of the limitations and complementarity of different knowledge systems

5.2 Ontology

- A multi-layered model of reality that accommodates both physical and metaphysical dimensions

- Recognition of the dynamic, process nature of existence

5.3 Ethics

- An expanded ethical framework based on cosmic interconnectedness

- Long-term, cosmos-wide perspective on moral responsibility

5.4 Psychology

- Understanding of the human psyche as a microcosm of cosmic principles

- Techniques for exploring and expanding consciousness as means of cosmic insight

5.5 Environmental Philosophy

- Deep ecological awareness based on fundamental interconnectedness

- Stewardship of Earth as a precious cradle of cosmic evolution

5.6 Social and Political Philosophy

- Recognition of unity amidst diversity in human societies

- Global perspective informed by cosmic citizenship

5.7 Art and Aesthetics

- Appreciation of beauty as a fundamental aspect of cosmic creativity

- Art as a means of exploring and expressing cosmic insights

6. Challenges and Criticisms

While a unified cosmic vision offers many benefits, it also faces several challenges and potential criticisms:

6.1 Scientific Reductionism

Scientific reductionism (SR) is a philosophical idea that explains complex systems by breaking them down into their simplest parts. It's based on the idea that knowledge of the complex can only be achieved through understanding its simpler components.

- Criticism that incorporating spiritual concepts dilutes scientific rigor

- Challenge of maintaining empirical grounding while embracing broader perspectives

6.2 Religious Orthodoxy

- Potential resistance from traditional religious viewpoints

- Challenge of honoring spiritual insights while embracing scientific discoveries

6.3 Cultural Appropriation

- Risk of superficial or inappropriate borrowing of concepts across cultures

- Need for deep, respectful engagement with diverse traditions

6.4 Logical Consistency

- Challenge of maintaining logical coherence when integrating diverse concepts

- Need for a robust philosophical framework to support integration

6.5 Practical Application

- Difficulty in translating cosmic insights into everyday life and decision-making

- Risk of detachment or escapism in face of immediate worldly challenges

6.6 Testability and Falsifiability

- Challenge of formulating testable hypotheses for some aspects of the unified vision

- Need to maintain scientific integrity while exploring broader concepts

7. Future Directions

As we look to the future, several promising directions emerge for further developing and applying this holistic cosmic vision:

7.1 Interdisciplinary Research

- Collaboration between physicists, neuroscientists, philosophers, and contemplatives

- Development of new research methodologies that integrate diverse ways of knowing

7.2 Consciousness Studies

- Advanced scientific investigation of altered states of consciousness

- Exploration of potential cosmic dimensions of consciousness

7.3 Ecological Applications

- Development of environmental policies and practices based on deep cosmic awareness

- Exploration of sustainable technologies inspired by cosmic principles

7.4 Educational Initiatives

- Creation of curricula that integrate scientific, philosophical, and contemplative approaches

- Training in multiple modes of knowing and cosmic perspectives

7.5 Artistic and Cultural Expressions

- New forms of art and media that express and explore cosmic insights

- Cultural movements that foster cosmic awareness and global unity

7.6 Technological Innovation

- Development of technologies that align with and enhance cosmic awareness

- Exploration of the ethical implications of advanced technologies from a cosmic perspective

7.7 Spiritual Evolution

- Integration of scientific insights into spiritual practices

- Evolution of spiritual traditions to encompass cosmic perspectives

8. Personal and Collective Transformation

Ultimately, a holistic cosmic vision is not merely an intellectual construct but a catalyst for personal and collective transformation:

8.1 Expanding Identity

- Shifting from a limited ego-based identity to a cosmic sense of self

- Recognizing oneself as an integral part of the cosmic process

8.2 Deepening Responsibility

- Embracing a sense of cosmic citizenship and responsibility

- Acting with awareness of the long-term, cosmic implications of our choices

8.3 Cultivating Wonder

- Maintaining a sense of awe and wonder at the beauty and mystery of the cosmos

- Approaching life with an attitude of continuous discovery and learning

8.4 Fostering Unity

- Recognizing the fundamental unity underlying human diversity

- Working towards global cooperation based on shared cosmic heritage

8.5 Evolving Consciousness

- Engaging in practices that expand and evolve consciousness

- Seeing personal growth as participation in cosmic evolution

9. Conclusion: A New Cosmic Story

As we conclude our exploration of cosmic origins and our attempt at synthesis, we find ourselves participants in the creation of a new cosmic story. This narrative weaves together the empirical discoveries of science, the profound insights of Hindu philosophy, and the contributions of other wisdom traditions into a tapestry that is richer and more comprehensive than any single perspective alone.

Our holistic cosmic vision sees the universe as a vast, interconnected whole, dynamically evolving yet grounded in a fundamental unity. It recognizes consciousness as intrinsic to the cosmos, seeing humans and other sentient beings as active participants in the unfolding of cosmic creativity. This vision acknowledges multiple layers of reality, from the quantum to the cosmic, from the physical to the metaphysical, each offering valid and complementary descriptions of the whole.

This new cosmic story offers a framework for addressing the big questions of existence – the nature of reality, the origin and destiny of the cosmos, the place of consciousness in the universe, and the meaning and purpose of life. It does so not by providing final answers, but by opening up new ways of exploring, understanding, and participating in the cosmic drama.

Importantly, this holistic vision is not a static endpoint but a dynamic, evolving understanding. It invites continuous exploration, refinement, and expansion. It calls us to be both humble students and active co-creators of the cosmic process.

As we move forward with this expanded cosmic perspective, we are challenged to live differently – with greater awareness, compassion, and responsibility. We are invited to see ourselves as integral parts of the cosmic whole, our actions rippling out across the vast ocean of space and time. We are called to be worthy ancestors to future generations and worthy descendants of our cosmic heritage.

In embracing this holistic cosmic vision, we open ourselves to a profound sense of meaning, connection, and purpose. We recognize that our individual stories are threads in the grand cosmic narrative, our personal journeys of growth and discovery mirroring the evolution of the cosmos itself.

As we close this chapter and this book, we stand in awe before the mystery and majesty of the cosmos. We are humbled by how much remains unknown and unknowable, yet inspired by how far our understanding has come. We look to the future with hope and determination, ready to continue the great adventure of cosmic exploration and self-discovery.

May this holistic cosmic vision serve as a guiding light, inspiring us to live with wisdom, compassion, and cosmic awareness. And may our ongoing exploration of cosmic origins, bridging the Big Bang and Brahman, contribute to the flourishing of life and consciousness in our precious corner of the cosmos and beyond.

Epilogue: The Ongoing Cosmic Journey

As we come to the close of our exploration into cosmic origins, bridging the scientific perspective of the Big Bang with the Hindu concept of Brahman, we find ourselves not at an endpoint, but at a new beginning. Our journey through the realms of modern cosmology and ancient wisdom has not provided us with final answers, but rather with a more expansive way of asking questions and a deeper appreciation for the profound mystery of existence.

We began this journey by peering into the earliest moments of our universe, witnessing in our mind's eye the explosive birth of space and time as described by the Big Bang theory. We marvelled at the precision of cosmic evolution, the delicate balance of forces that allowed for the formation of galaxies, stars, and eventually, life itself. The scientific narrative revealed a cosmos of staggering vastness and complexity, evolving over billions of years from primal simplicity to the rich tapestry of existence we observe today.

Alongside this, we delved into the profound insights of Hindu philosophy, particularly the concept of Brahman – the ultimate, unchanging reality that underlies and permeates all of existence. We explored how this ancient wisdom tradition views the cosmos not as a mere physical entity, but as a manifestation of consciousness itself. We contemplated the ideas of maya (cosmic illusion), the cycles of creation and dissolution, and the ultimate unity of all being.

As we brought these perspectives into dialogue, we discovered intriguing parallels and points of convergence. Both the Big Bang theory and the concept of Brahman point towards a fundamental unity as the source of all existence. Both suggest a progression from simplicity to complexity. Both lead us to question our ordinary perception of reality, revealing a cosmos far stranger and more mysterious than our everyday experience suggests.

Yet, we also encountered significant differences and challenges in bridging these worldviews. The empirical methodology of science and the intuitive insights of spiritual traditions often seem to speak different languages and operate on different planes of understanding. We grappled with questions of consciousness, the nature of time, the role of the observer, and the limits of human knowledge.

Through this exploration, a more holistic vision began to emerge – one that honors both the rigorous findings of science and the profound insights of spiritual wisdom. This vision sees the universe as a vast, interconnected whole, dynamically evolving yet grounded in a fundamental unity. It recognizes consciousness not as a mere byproduct of material processes, but as an intrinsic aspect of the cosmos. It acknowledges multiple layers of reality, from the quantum to the cosmic, from the physical to the metaphysical, each offering valid and complementary descriptions of the whole.

This emerging cosmic vision challenges us to expand our understanding of ourselves and our place in the universe. It invites us to see ourselves not as separate observers of the cosmos, but as active participants in its unfolding. We are, in a very real sense, the universe becoming aware of itself, the means by which the cosmos contemplates its own existence.

As we stand here at the frontier of human knowledge, looking out into the vast mystery of the cosmos, we are filled with a profound sense of humility and awe. The journey of discovery that brought us to this point – from the first curious glances at the night sky to our current ability to probe the earliest moments of cosmic history – is itself a testament to the remarkable capacities of human consciousness.

Yet, we must acknowledge that our understanding, however far it has come, remains limited. The ultimate nature of reality, the deepest truths of existence, may always elude complete human comprehension. This realization, rather than being a source of frustration, can be seen as an invitation – an endless call to deeper exploration, to expanded awareness, to continued growth and evolution.

As we move forward from here, carrying the insights gained from this bridging of perspectives, several pathways of further exploration open before us:

1. Continued Scientific Inquiry: The quest to understand our cosmic origins is far from over. Ongoing research in cosmology, quantum physics, and related fields continues to push the boundaries of our knowledge. Future discoveries may shed new light on the earliest moments of the universe, the nature of dark matter and dark energy, and the possibility of multiple universes.

2. Deeper Philosophical Integration: The dialogue between scientific and spiritual perspectives on cosmic origins is still in its early stages. There is rich potential for further philosophical work that can help to create more comprehensive frameworks for understanding reality.

3. Consciousness Studies: The exploration of consciousness – its nature, its role in the cosmos, and its potential for expansion – emerges as a key area for future research. This field may serve as a bridge between scientific and spiritual approaches to understanding reality.

4. Ethical and Practical Applications: Our expanded cosmic perspective has profound implications for how we live our lives and organize our societies. There is important work to be done in translating these cosmic insights into practical ethics, environmental stewardship, and social organization.

5. Personal and Collective Transformation: Perhaps most importantly, the insights gained from this cosmic exploration invite us to embark on journeys of personal and collective transformation. How can we live in greater alignment with cosmic principles? How can we expand our awareness to embrace more of the cosmic whole?

As we close this book and look to the future, we are reminded that our exploration of cosmic origins is not merely an intellectual exercise. It is a vital part of humanity's ongoing journey of self-discovery and evolution. In seeking to understand the birth and nature of the cosmos, we are ultimately seeking to understand ourselves – our origins, our nature, our potential, and our purpose.

The bridge we have built between the Big Bang and Brahman is not a final structure, but a starting point for further exploration. It is an invitation to continue the great adventure of cosmic discovery, to remain open to new insights from both scientific inquiry and spiritual wisdom, and to allow our understanding of the cosmos to transform our way of being in the world.

As you leave these pages and return to your daily life, carry with you the vastness of cosmic time and space, the profound interconnectedness of all existence, and the mystery that lies at the heart of reality. Let these cosmic perspectives inform your choices, inspire your actions, and deepen your appreciation for the miraculous gift of existence.

For we are all, in our essence, children of the cosmos – whether understood through the lens of the Big Bang or Brahman. Our individual stories are threads in the grand cosmic narrative, our personal journeys of growth and discovery mirroring the evolution of the universe itself.

May your ongoing exploration of cosmic origins be a source of wonder, wisdom, and awakening. And may it contribute, in whatever way possible, to the flourishing of life and consciousness in our precious corner of the cosmos and beyond.

The journey continues. The cosmos awaits.

Appendix A

SWOT (Strength – Weakness- Opportunity- Threat) analysis of the Two domains

With the attempt in this book to synthesize the two approaches to understand our world view having been completed we are in a position to attempt evaluation of comparative strengths and weaknesses of the two approaches and also speculate on the possible opportunities and threats in likely future.

This SWOT analysis may provide us a perspective on how to navigate our journey towards understanding the reality better. Both approaches are complementary and we possibly cannot leave one for the other in any holistic search for our existential questions. While empirical approach is having an external focus using our five senses and analytical ability of mind-intellect, Vedantic approach demands that we turn inward to learn deeper truths through experiential knowledge and in this inward journey, our conventional sources of knowledge, our five senses and mind-intellect duo are to be suspended.

In the following paragraphs we go through this exercise using the insights gained in previous chapters.

Empirical Approach of Modern Cosmology

Strengths

1. Based on empirical evidence and scientific method

- Example: The discovery of cosmic microwave background radiation in 1964, which provided strong evidence for the Big Bang theory.

2. Utilizes advanced technology and mathematical models

- Example: The use of the Hubble Space Telescope to observe distant galaxies and measure the expansion rate of the universe.

3. Constantly evolving with new discoveries

- Example: The 2015 detection of gravitational waves by LIGO, confirming a key prediction of Einstein's general relativity.

4. Provides testable predictions and hypotheses

- Example: The prediction and subsequent discovery of the Higgs boson at CERN in 2012.

5. Widely accepted in the scientific community

- Example: The broad consensus on the Big Bang theory and the expansion of the universe among cosmologists and astrophysicists.

Weaknesses

1. Limited by current technological capabilities

- Example: The inability to directly observe dark matter, despite its apparent gravitational effects.

2. Cannot yet explain certain phenomena

- Example: The nature of dark energy, which is thought to be driving the accelerating expansion of the universe, remains a mystery.

3. May not address existential or philosophical questions

- Example: Cosmology doesn't provide answers to questions like "Why does the universe exist?" or "What is the purpose of life?"

4. Theories can be complex and difficult for the general public to understand

- Example: The concept of multiple dimensions in string theory is mathematically sound but challenging for non-experts to grasp.

5. Subject to paradigm shifts as new evidence emerges

- Example: The discovery of the accelerating expansion of the universe in 1998 led to a major revision of cosmological models.

Opportunities

1. Potential for groundbreaking discoveries about the universe

- Example: The possibility of detecting primordial gravitational waves, which could provide evidence for cosmic inflation.

2. Advancement of space exploration and technologies

- Example: The development of more powerful telescopes like the James Webb Space Telescope, enabling observation of the earliest galaxies.

3. Interdisciplinary collaborations

- Example: The intersection of cosmology and particle physics in studying the early universe and fundamental forces.

4. Public engagement and education in science

- Example: Popular science books and documentaries about cosmology, like "A Brief History of Time" by Stephen Hawking.

5. Development of new technologies with practical applications

- Example: Spin-off technologies from space research, such as improved medical imaging techniques.

Threats

1. Funding challenges for large-scale research projects

- Example: Budget cuts affecting major projects like the Superconducting Super Collider, which was cancelled in 1993.

2. Potential misinterpretation or misuse of findings

- Example: Misunderstanding of the term "theory" in scientific context, leading to dismissal of well-established ideas like evolution.

3. Resistance from some religious or philosophical groups

- Example: Opposition to the Big Bang theory from certain creationist groups.

4. Overspecialisation leading to communication barriers

- Example: Difficulty in translating highly technical cosmological concepts for public understanding or interdisciplinary collaboration.

5. Ethical concerns about certain research directions

- Example: Debates about the safety of creating miniature black holes in particle accelerators like the Large Hadron Collider (LHC).

Advaita Vedanta

Strengths

1. Provides a comprehensive philosophical framework

- Example: The concept of Brahman as the ultimate, non-dual reality underlying all existence.

2. Addresses existential questions and the nature of consciousness

- Example: The exploration of the nature of the self (Atman) and its relationship to universal consciousness.

3. Has a long historical and cultural tradition

- Example: Texts like the Upanishads and works by philosophers like Adi Shankara, dating back over a thousand years.

4. Offers practical methods for self-realization

- Example: Meditation techniques and self-inquiry practices to realize one's true nature.

5. Promotes unity and interconnectedness of all things

- Example: The teaching that all diversity is ultimately an expression of one underlying reality (Brahman).

Weaknesses

1. Lacks empirical evidence by scientific standards

- Example: The claim of non-dual consciousness is not directly testable through current scientific methods.

2. Relies heavily on interpretation of ancient texts

- Example: Different interpretations of key concepts in the Upanishads can lead to varying schools of thought.

3. Can be abstract and difficult to grasp for some

- Example: The concept of maya (illusion) and its role in perception of the world can be challenging to understand.

4. May not address specific, practical problems in the physical world

- Example: Advaita Vedanta doesn't provide direct solutions to issues like climate change or economic inequality.

5. Subjective experiences are hard to verify or replicate

- Example: Mystical experiences reported during deep meditation cannot be objectively measured or reproduced.

Opportunities

1. Integration with modern psychology and neuroscience

- Example: Studies on the effects of meditation on brain structure and function, bridging ancient practices with modern science.

2. Potential applications in mental health and well-being

- Example: Mindfulness-based therapies inspired by Vedantic concepts for treating depression and anxiety.

3. Cross-cultural dialogue and understanding

- Example: Conferences and forums bringing together Eastern and Western philosophers to discuss consciousness and reality.

4. Inspiration for new approaches in theoretical physics

- Example: Parallels drawn between Advaita Vedanta's non-dualism and certain interpretations of quantum mechanics.

5. Development of ethical frameworks based on non-dualism

- Example: Environmental ethics derived from the idea of interconnectedness of all beings.

Threats

1. Misinterpretation or commercialization of teachings

- Example: Superficial adoption of Vedantic concepts in "New Age" movements, potentially diluting or misrepresenting core teachings.

2. Potential conflict with other religious or philosophical views

- Example: Disagreements with dualistic religious traditions over the nature of God and the self.

3. Marginalization in predominantly materialistic societies

- Example: Difficulty in gaining acceptance or funding for Vedantic studies in academic institutions focused on empirical research.

4. Loss of traditional knowledge and practices

- Example: Decline in the number of scholars proficient in Sanskrit, the language of many original Advaita Vedanta texts.

5. Challenges in maintaining relevance in a rapidly changing world

- Example: Adapting ancient teachings to address modern ethical dilemmas posed by eerging technologies.

Appendix B

Glossary of Modern Cosmology Terms

A

- Absolute magnitude: The brightness of a celestial object as it would appear if it were 10 parsecs (about 32.6 light-years) away from Earth.

- Accelerating universe: The observation that the expansion of the universe is speeding up, likely due to dark energy.

- Active galactic nucleus (AGN): The compact region at the center of a galaxy that has a much higher than normal luminosity over some or all of the electromagnetic spectrum.

B

- Baryon: A subatomic particle made up of three quarks, such as protons and neutrons.

- Big Bang: The prevailing cosmological model for the observable universe from the earliest known periods through its subsequent large-scale evolution.

- Black hole: A region of spacetime exhibiting gravitational acceleration so strong that nothing—no particles or even electromagnetic radiation such as light—can escape from it.

C

- Cosmic inflation: A theory of exponential expansion of space in the early universe.

- Cosmic microwave background (CMB): Electromagnetic radiation as a remnant from an early stage of the universe in Big Bang cosmology.

- Cosmological constant: A term added by Einstein to his general relativity equations to achieve a stationary universe.

D

- Dark energy: An unknown form of energy which is hypothesized to permeate all of space, tending to accelerate the expansion of the universe.

- Dark matter: A hypothetical form of matter that is thought to account for approximately 85% of the matter in the universe and about a quarter of its total energy density.

- Doppler effect: The change in frequency of a wave for an observer moving relative to its source.

E

- Event horizon: A boundary in spacetime beyond which events cannot affect an outside observer.

- Exoplanet: A planet that orbits a star other than the Sun.

F

- Friedmann equations: The equations of motion governing the expansion of space in homogeneous and isotropic models of the universe within the context of general relativity.

G

- Galaxy: A gravitationally bound system of stars, stellar remnants, interstellar gas, dust, and dark matter.

- Gravitational lensing: The bending of light by massive objects in the universe, as explained by Einstein's theory of general relativity.

H

- Hubble constant: A measure of the current expansion rate of the universe.

- Hubble's law: The observation that galaxies are moving away from the Earth at speeds proportional to their distance.

I

- Inflation: A theory of exponential expansion of space in the early universe.

- Interstellar medium: The matter and radiation that exists in the space between the star systems in a galaxy.

L

- Lambda-CDM model: The current standard model of Big Bang cosmology, which includes cold dark matter and a cosmological constant (Λ).

- Light-year: The distance that light travels in one Earth year; approximately 9.46 trillion kilometers.

M

- Multiverse: A hypothetical group of multiple universes.

N

- Nebula: An interstellar cloud of dust, hydrogen, helium and other ionized gases.

- Neutron star: The collapsed core of a large star which before collapse had a total mass of between 10 and 29 solar masses.

O

- Observable universe: The portion of the universe that can be seen from Earth.

P

- Parallax: The apparent displacement of an observed object due to a change in the observer's point of view.

- Parsec: A unit of length used to measure large distances to astronomical objects, equal to about 3.26 light-years.

Q

- Quasar: An extremely luminous active galactic nucleus.

R

- Redshift: The increase in wavelength of electromagnetic radiation as it travels through expanding space.

S

- Singularity: A point where a physical model predicts infinite density and zero volume.

- Standard candle: An astronomical object that has a known absolute luminosity.

- Supernova: A powerful and luminous stellar explosion.

T

- Thermal equilibrium: A state in which a system is in thermal balance with its surroundings.

W

- White dwarf: A stellar core remnant composed mostly of electron-degenerate matter.

Z

- Z factor: Another term for redshift, denoted by the letter z.

Appendix C

Glossary of Hindu Vedanta Terms

A

- Advaita: Non-dualism; the doctrine that reality is fundamentally one, without a second.
- Aham Brahmasmi: "I am Brahman"; one of the Mahavakyas (great sayings) of the Upanishads.
- Ananda: Bliss; one of the three aspects of Brahman (Sat-Chit-Ananda).
- Atman: The individual self or soul; in Advaita Vedanta, identical with Brahman.
- Avidya: Ignorance; the misconception of reality that leads to suffering.

B

- Bhakti: Devotion to God; one of the paths to liberation.
- Brahman: The ultimate reality; the one, eternal, and infinite cosmic principle.
- Buddhi: Intellect; the faculty of discrimination and decision-making.

C

- Chit: Consciousness; one of the three aspects of Brahman (Sat-Chit-Ananda).

D

- Dvaita: Dualism; the philosophy that differentiates between the individual soul and Brahman.
- Dharma: Righteous living; duty; cosmic order.

G

- Guna: Quality or attribute; specifically refers to the three qualities of nature (sattva, rajas, tamas).
- Guru: A spiritual teacher or guide.

J

- Jiva: The individual soul, bound by karma and subject to reincarnation.
- Jnana: Knowledge; especially spiritual knowledge leading to liberation.

K

- Karma: Action and its consequences; the law of cause and effect.
- Kosha: Sheath or layer; refers to the five sheaths covering the Atman.

M

- Maya: The illusory nature of the phenomenal world; the power that veils the true nature of reality.
- Moksha: Liberation from the cycle of birth and death; the ultimate goal of Vedanta.

N

- Neti Neti: "Not this, not this"; a method of negation used to describe Brahman.
- Nirguna Brahman: Brahman without attributes or qualities.

O

- Om: The primordial sound; a sacred syllable representing Brahman.

P

- Prakriti: Primordial nature; the material cause of the universe.
- Prana: Life force or vital energy.
- Purusha: The eternal, unchanging principle of consciousness.

R

- Rishi: A sage or seer who has attained self-realization.

S

- Saguna Brahman: Brahman with attributes or qualities.
- Samsara: The cycle of birth, death, and rebirth.
- Sannyasa: Renunciation; the fourth stage of life in Hinduism.
- Sat: Existence; one of the three aspects of Brahman (Sat-Chit-Ananda).
- Satya: Truth; ultimate reality.
- Shruti: That which is heard; refers to the Vedas and Upanishads.
- Smriti: That which is remembered; refers to secondary scriptures like the Bhagavad Gita.

T

- Tattva: Principle or element; fundamental aspects of reality.
- Turiya: The fourth state of consciousness, beyond waking, dreaming, and deep sleep.

U

- Upanishad: Philosophical texts that form the theoretical basis
for the Hindu religion.

V

- Vasana: Subtle desire or mental impression from past lives.
- Vedanta: "End of the Vedas"; one of the six orthodox schools
of Hindu philosophy.
- Viveka: Discrimination between the real and the unreal.

Y

- Yoga: Union; refers to various spiritual and ascetic practices.

Appendix D

Conceptual Mapping: Modern Cosmology and Vedanta Cosmology

1. Fundamental Reality

- **Modern Cosmology**: The physical universe, governed by natural laws

- **Vedanta: Brahman** (ultimate reality)

Possible connection: Both systems seek to describe the fundamental nature of reality, though they approach it from different perspectives

2. Origin of the Universe

- **Modern Cosmology**: Big Bang theory

- **Vedanta**: Cycles of creation and dissolution (Brahma's day and night)

Possible connection: Both systems acknowledge a beginning of the current state of the universe, though Vedanta sees this as cyclical rather than a single event.

3.Fundamental Particles

- **Modern Cosmology**: Quarks, leptons, bosons

- **Vedanta:** Prakriti (primordial nature) composed of three gunas (sattva, rajas, tamas)

Possible connection: Both systems attempt to describe the fundamental building blocks of reality, though in very different ways.

4.Dark Energy/Expanding Universe

- **Modern Cosmology**: Dark energy causing accelerating expansion

- **Vedanta**: Maya (the illusive power that creates the appearance of the physical universe)

Possible connection: Both concepts relate to unseen forces shaping the observable universe.

5.Consciousness

- **Modern Cosmology**: Emergent property of complex neural systems

- **Vedanta**: Fundamental aspect of reality (Chit - one of the three aspects of Brahman)

Possible connection: Both acknowledge consciousness, but with vastly different interpretations of its nature and origin.

6.Multiple Universes

- **Modern Cosmology**: Multiverse theory

- **Vedanta**: Multiple lokas (planes of existence)

Possible connection: Both entertain the possibility of realities beyond our observable universe.

7.Nature of Time

- **Modern Cosmology**: Fourth dimension in spacetime

- **Vedanta**: Maya's power of illusion, creating the appearance of sequential events

Possible connection: Both systems recognize time as a fundamental aspect of reality, though they conceptualize it differently.

8.Evolution of the Universe

- **Modern Cosmology**: Cosmic evolution through physical processes

- **Vedanta**: Gradual unfoldment of divine consciousness

Possible connection: Both describe a process of change and development in the cosmos, albeit through very different mechanisms.

9.Observer Effect

- **Modern Cosmology**: Quantum observer effect

- **Vedanta**: The role of consciousness in shaping perceived reality

Possible connection: Both systems acknowledge a relationship between the observer and the observed, though they interpret this relationship differently.

10.Ultimate Fate of the Universe

- **Modern Cosmology**: Various theories (Big Freeze, Big Crunch, Big Rip)

- **Vedanta**: Pralaya (cosmic dissolution) followed by new creation

Possible connection: Both consider long-term cosmic scenarios, though Vedanta sees this as part of an eternal cycle.

11.Underlying Unity

- **Modern Cosmology**: Search for a unified theory of everything

- **Vedanta**: Non-dualism (Advaita) - all is one Brahman

Possible connection: Both seek to understand the underlying unity of diverse phenomena, though through very different approaches.

Appendix E

Challenges and Limitations in Comparing Modern and Vedanta Cosmologies

Key Challenges and Limitations

1. Epistemological Differences:

- Modern cosmology relies on empirical observation, mathematical models, and the scientific method.

- Vedanta is based on scriptural authority, philosophical reasoning, and mystical experience.

2. Purpose and Scope:

- Modern cosmology aims to describe the physical universe and its laws.

- Vedanta seeks to understand the nature of reality and the self, with cosmology as part of a broader spiritual framework.

3. Cultural and Historical Context:

- Modern cosmology emerged from Western scientific tradition.

- Vedanta developed within the cultural and philosophical context of ancient India.

4. Language and Terminology:

- Scientific terms have precise, operationalized definitions.

- Vedantic concepts often have multiple layers of meaning and can be interpreted differently.

5. Metaphysical Assumptions:

- Modern cosmology generally assumes methodological naturalism.

- Vedanta includes non-physical, transcendent aspects of reality.

6. Time Scales and Cyclicity:

- Modern cosmology deals with billions of years but generally sees time as linear.

- Vedanta often conceptualizes time in vast cycles (yugas and kalpas).

7. Role of Consciousness:

- In modern cosmology, consciousness is generally not a fundamental feature of the universe.

- In Vedanta, consciousness (Chit) is a fundamental aspect of reality.

Expert Perspectives

1. Dr. Karan Singh, scholar of Vedanta and former Indian diplomat:

"While there are fascinating parallels between modern cosmology and Vedantic concepts, we must be cautious not to force-fit ancient wisdom into modern scientific frameworks. The underlying paradigms and purposes are quite different."

2. Dr. Fritjof Capra, physicist and systems theorist:

"The parallels between modern physics and Eastern mysticism are striking. However, they are analogies, not direct equivalences. The danger lies in oversimplifying either tradition in an attempt to harmonize them."

3. Dr. V.V. Raman, physicist and philosopher of science:

"Comparisons between modern science and ancient worldviews can be intellectually stimulating, but we must remember that they arise from very different epistemological foundations. What looks like a similarity may have very different implications in each system."

4. Dr. B.N. Faruqui, astrophysicist:

"While both modern cosmology and Vedanta grapple with fundamental questions about the nature of reality, their methodologies and goals are distinct. Modern cosmology seeks testable explanations for observed phenomena, while Vedanta aims at spiritual realization."

5. Dr. Anindita Balslev, philosopher and cross-cultural thinker:

"The dialogue between modern science and Vedanta can be enriching, but it requires careful navigation. We must avoid the temptation to validate ancient ideas with modern science or vice versa. Instead, we should appreciate each tradition on its own terms while remaining open to mutual insights."

6. Dr. Pankaj Jain, professor of philosophy and religion:

"When comparing Vedanta and modern cosmology, we must be aware of the risk of decontextualization. Vedantic concepts are part of a holistic philosophical and spiritual system, and isolating cosmological ideas from this context can lead to misinterpretation."

Appendix F

Similarities between Quantum Measurement and Vedantic Concepts

1. Observer-Dependent Reality

Quantum Perspective:

In quantum mechanics, the act of observation seems to play a crucial role in determining the state of a system. The wave function collapse, which occurs upon measurement, suggests that reality in some sense depends on the observer.

Vedantic Perspective:

In Advaita Vedanta, the observed world (jagat) is considered to be an emanation or appearance (vivarta) of Brahman, the ultimate reality. This world is seen as dependent on consciousness (Brahman) for its apparent existence.

Similarity:

Both perspectives suggest a deep connection between the observer (or consciousness) and the observed reality, challenging the notion of an objective, observer-independent world.

2. Superposition and Non-duality

Quantum Perspective:

Quantum systems can exist in superposition states, where they seem to occupy multiple states simultaneously until observed.

Vedantic Perspective:

Advaita Vedanta posits the concept of non-duality (advaita), where the apparent multiplicity of the world is ultimately an illusion (maya), and only Brahman truly exists.

Similarity:

Some researchers see a parallel between quantum superposition and the Vedantic idea that the diverse world of experience emerges from a unified, undifferentiated reality.

3. Interconnectedness

Quantum Perspective:

Quantum entanglement demonstrates that particles can be interconnected in ways that defy classical physics, suggesting a fundamental interconnectedness in nature.

Vedantic Perspective:

Vedanta teaches that all things are interconnected, being manifestations of the same ultimate reality (Brahman).

Similarity:

Both views challenge the classical notion of separate, independent entities and point to a deeper, underlying unity.

4. Limits of Measurement and Knowledge

Quantum Perspective:

Heisenberg's uncertainty principle sets fundamental limits on our ability to simultaneously measure certain pairs of physical properties.

Vedantic Perspective:

Vedanta recognizes limits to conceptual knowledge and emphasizes direct experience (anubhava) for understanding ultimate reality.

Similarity:

Both perspectives acknowledge inherent limitations in our ability to fully measure or conceptualize reality through ordinary means.

5. Role of Consciousness

Quantum Perspective:

Some interpretations of quantum mechanics, particularly those influenced by John von Neumann and Eugene Wigner, have suggested a special role for consciousness in the measurement process.

Vedantic Perspective:

Consciousness (often equated with Brahman) is central to Vedantic philosophy, being seen as the fundamental substrate of all existence.

Similarity:

Both views potentially assign a primary role to consciousness in the nature of reality, though this is more universally accepted in Vedanta than in quantum physics.

6. Critiques and Cautions

While these similarities are intriguing, many scientists and philosophers urge caution in drawing too close a parallel:

1. **Different domains**: Quantum mechanics is a scientific theory dealing with the microscopic world, while Vedanta is a philosophical and spiritual system addressing the nature of reality and self.

2. **Rigor of approach**: Quantum mechanics relies on mathematical formalism and experimental verification, whereas Vedantic insights are often based on introspection and spiritual experience.

3. **Risk of misinterpretation**: There's a danger of oversimplifying or misinterpreting both quantum mechanics and Vedantic philosophy when drawing these parallels.

4. **Cultural and historical context**: These ideas developed in very different cultural and historical contexts, which should be considered when comparing them.

Conclusion

While there are intriguing similarities between some aspects of quantum mechanics and Vedantic philosophy, it's important to approach these comparisons with nuance and critical thinking. The parallels may offer interesting philosophical insights, but they should not be taken as direct equivalences between scientific theory and spiritual philosophy.

225

Appendix G

Quantum Entanglement and Vedantic Unity in Brahman

Quantum Entanglement

Quantum entanglement is a phenomenon in quantum physics where two or more particles become correlated in such a way that the quantum state of each particle cannot be described independently of the others, even when separated by large distances.

Key features:

1. Non-local correlations

2. Instantaneous influence (seemingly faster than light)

3. Holistic nature of quantum systems

Vedantic Concept of Unity in Brahman

In Advaita Vedanta, Brahman is considered the ultimate reality, the underlying unity from which all apparent diversity emerges.

Key features:

1. Non-dual nature of reality

2. Interconnectedness of all things

3. Transcendence of space and time

Perceived Equivalences

Some philosophers and researchers have drawn parallels between these concepts:

1. Holistic Nature of Reality

  - Entanglement: The state of an entangled system must be described as a whole, not as independent parts.

  - Vedanta: All existence is seen as manifestations of the single, undivided Brahman.

2. Transcendence of Space-Time

  - Entanglement: Entangled particles maintain their correlation regardless of distance, seemingly instantaneously.

  - Vedanta: Brahman is described as beyond space and time, omnipresent and eternal.

3. Underlying Unity

  - Entanglement: Suggests a deep, underlying connection between particles that persists despite apparent separation.

  - Vedanta: Posits that all apparent diversity is ultimately grounded in the unity of Brahman.

4. Non-locality

  - Entanglement: Exhibits non-local correlations that seem to defy classical notions of locality.

- Vedanta: Describes Brahman as all-pervading, transcending spatial limitations.

5. **Observer Effect**

- Entanglement: Measurement of one entangled particle instantly affects its partner, reminiscent of the observer effect in quantum mechanics.

- Vedanta: Emphasizes the role of consciousness (often equated with Brahman) in the manifestation of the perceived world.

Philosophical Interpretations

1. **Interconnectedness**: Both concepts suggest a fundamental interconnectedness of all things, challenging the reductionist view of reality.

2. **Limits of Perception**: Entanglement and Vedantic philosophy both point to limitations in our ordinary perception of reality.

3. **Wholeness**: Both ideas emphasize the importance of considering systems as wholes rather than just collections of parts.

4. **Consciousness and Reality**: Some interpretations of both quantum mechanics and Vedanta suggest a deep relationship between consciousness and the nature of reality.

Cautions and Criticisms

1. **Different Domains**: Quantum entanglement is a specific physical phenomenon, while Vedantic unity is a metaphysical concept.

2. **Scientific vs Philosophical**: Entanglement is subject to scientific experimentation and mathematical formulation, whereas Vedantic concepts are philosophical and experiential.

3. **Scale Differences**: Quantum effects like entanglement typically occur at microscopic scales, while Vedantic unity is proposed as a universal principle.

4. **Risk of Overinterpretation**: There's a danger of stretching analogies too far and misinterpreting both quantum mechanics and Vedantic philosophy.

5. **Cultural Context**: These ideas developed in vastly different cultural and historical contexts, which should be considered in any comparison.

Conclusion

While there are intriguing parallels between quantum entanglement and the Vedantic concept of unity in Brahman, it's crucial to approach these comparisons with caution. The similarities may offer valuable philosophical insights and inspire new ways of thinking about reality, but they should not be taken as direct scientific equivalences. Both quantum mechanics and Vedantic philosophy are complex fields with their own rigorous methodologies and contexts, which must be respected in any comparative analysis.

Bibliography

Aspect, A., Dalibard, J., & Roger, G. (1982). Experimental Test of Bell's Inequalities Using Time-Varying Analyzers. Physical Review Letters, 49(25), 1804-1807.

Balslev, A. N. (1983). A Study of Time in Indian Philosophy. Otto Harrassowitz.

Barbour, J. (1999). The End of Time: The Next Revolution in Physics. Oxford University Press.

Barrow, J. D. (2011). The Book of Universes: Exploring the Limits of the Cosmos. W. W. Norton & Company.

Barrow, J. D., & Tipler, F. J. (1986). The Anthropic Cosmological Principle. Oxford University Press.

Bhavasar, S. N., & Boyer, R. W. (2009). Major progress linking modern science and Vedic science.

Bohm, D. (1980). Wholeness and the Implicate Order. Routledge..

Capra, F. (1975). The Tao of Physics: An Exploration of the Parallels between Modern Physics and Eastern Mysticism. Shambhala Publications.

Capra, F. (2010). The Tao of Physics: An Exploration of the Parallels between Modern Physics and Eastern Mysticism. Shambhala.

Capra, F., & Luisi, P. L. (2014). The Systems View of Life: A Unifying Vision. Cambridge University Press.

Carroll, S. (2010). From Eternity to Here: The Quest for the Ultimate Theory of Time. Dutton.

Carroll, S. (2019). Something Deeply Hidden: Quantum Worlds and the Emergence of Spacetime. Dutton.

Carroll, S. M. (2010). From Eternity to Here: The Quest for the Ultimate Theory of Time. Dutton.

Carter, B. (1974). Large Number Coincidences and the Anthropic Principle in Cosmology. In M. S. Longair (Ed.), Confrontation of Cosmological Theories with Observational Data. Reidel.

Chalise, K. R. (2021). The Science of Religion in the Vedic Texts: A Physico-Theological Perspective. The Outlook: Journal of English Studies, 72-82.

Chalmers, D. J. (1996). The Conscious Mind: In Search of a Fundamental Theory. Oxford University Press.

Chandrasekharayya, S., & Vanaja, M. (2014). The Vedantic Concept of Consciousness vis-à-vis Quantum Mechanics. International Journal of Scientific and Research Publications, 4(5), 1-5.

Chapple, C. K. (2001). Hinduism and Ecology: The Intersection of Earth, Sky, and Water. Harvard University Press.

Chapple, C. K. (Ed.). (2004). Jainism and Ecology: Nonviolence in the Web of Life. Harvard University Press.

Chatterjee, S. C. (1950). The Fundamentals of Hinduism: A Philosophical Study. Das Gupta & Co.

Chopra, D. (1989). Quantum Healing: Exploring the Frontiers of Mind/Body Medicine. Bantam Books.

Chopra, D., & Kafatos, M. (2017). You Are the Universe: Discovering Your Cosmic Self and Why It Matters. Harmony.

Clooney, F.X. (1993). Theology After Vedanta: An Experiment in Comparative Theology. State University of New York Press.

Coomaraswamy, A. K. (1947). Time and Eternity. Artibus Asiae Publishers.

Coyne, G. V., & Omizzolo, A. (2002). Wayfarers in the Cosmos: The Human Quest for Meaning. Crossroad.

d'Espagnat, B. (1979). The Quantum Theory and Reality. Scientific American, 241(5), 158- 181.

Dalai Lama. (2005). The Universe in a Single Atom: The Convergence of Science and Spirituality. Morgan Road Books.

Dasgupta, S. (1922). A History of Indian Philosophy. Cambridge University Press.

Dasgupta, S. (2010). Science and Philosophy in the Indian Buddhist Classics, Vol. 1: The Physical World. Wisdom Publications.

Datta, D. P. (2013). A new interpretation of the role of entropy in macroscopic systems. Physica A: Statistical Mechanics and its Applications, 392(17), 3454-3464.

Davies, P. (1995). About Time: Einstein's Unfinished Revolution. Simon & Schuster.

Davies, P. (2006). The Goldilocks Enigma: Why Is the Universe Just Right for Life? Allen Lane.

Dehaene, S. (2014). Consciousness and the Brain: Deciphering How the Brain Codes Our Thoughts. Viking.

Deussen, P. (1906). The Philosophy of the Upanishads. T. & T. Clark.

Deutsch, D. (1997). The Fabric of Reality: The Science of Parallel Universes and Its Implications. Allen Lane.

Deutsch, D. (1997). The Fabric of Reality: The Science of Parallel Universes and Its Implications. Allen Lane.

Deutsch, D. (2011). The Beginning of Infinity: Explanations That Transform the World. Allen Lane.

Deutsch, E. (1969). Advaita Vedanta: A Philosophical Reconstruction. University of Hawaii Press.

Dobson, J. (1983). Advaita Vedanta and Modern Science. Routledge.

Eliade, M. (1954). The Myth of the Eternal Return: Cosmos and History. Princeton University Press.

Feynman, R. P. (1965). The Character of Physical Law. MIT Press.

Feynman, R. P. (1985). QED: The Strange Theory of Light and Matter. Princeton University Press.

Flood, G. (1996). An Introduction to Hinduism. Cambridge University Press.

Fort, A.O. (1998). Jivanmukti in Transformation: Embodied Liberation in Advaita and Neo-Vedanta. State University of New York Press.

Fraser, J. T. (1987). Time, the Familiar Stranger. University of Massachusetts Press.

Gödel, K. (1931). Über formal unentscheidbare Sätze der Principia Mathematica und verwandter Systeme I. Monatshefte für Mathematik und Physik, 38(1), 173-198.

Gödel, K. (1949). An Example of a New Type of Cosmological Solutions of Einstein's Field Equations of Gravitation. Reviews of Modern Physics, 21(3), 447-450.

Goswami, A. (1993). The Self-Aware Universe: How Consciousness Creates the Material World. Tarcher/Putnam.

Goswami, A. (1995). The Self-Aware Universe: How Consciousness Creates the Material World. Tarcher.

Goswami, A. (2001). Physics of the Soul: The Quantum Book of Living, Dying, Reincarnation, and Immortality. Hampton Roads Publishing.

Greene, B. (1999). The Elegant Universe: Superstrings, Hidden Dimensions, and the Quest for the Ultimate Theory. W.W. Norton.

Greene, B. (2004). The Fabric of the Cosmos: Space, Time, and the Texture of Reality. Alfred A. Knopf.

Greene, B. (2020). Until the End of Time: Mind, Matter, and Our Search for Meaning in an Evolving Universe. Alfred A. Knopf.

Gupta, B. (1998). The Disinterested Witness: A Fragment of Advaita Vedanta Phenomenology. Northwestern University Press.

Guth, A. H. (1997). The Inflationary Universe: The Quest for a New Theory of Cosmic Origins. Perseus Books.

Hagelin, J. S. (1987). Is consciousness the unified field? A field theorist's perspective. Modern Science and Vedic Science, 1(1), 29-87.

Haisch, B. (2006). The God Theory: Universes, Zero-Point Fields and What's Behind It All. Red Wheel/Weiser.

Halbfass, W. (1988). India and Europe: An Essay in Understanding. State University of New York Press.

Hameroff, S., & Penrose, R. (2014). Consciousness in the universe: A review of the 'Orch OR' theory. Physics of Life Reviews, 11(1), 39-78.

Hawking, S. W. (1988). A Brief History of Time: From the Big Bang to Black Holes. Bantam Books.

Hawking, S. W., & Mlodinow, L. (2010). The Grand Design. Bantam.

Heisenberg, W. (1958). Physics and Philosophy: The Revolution in Modern Science. Harper & Brothers.

Herbert, N. (1985). Quantum Reality: Beyond the New Physics. Anchor Books.

Hiriyanna, M. (1932). Outlines of Indian Philosophy. London: George Allen & Unwin.

Holman, M. (2013). Scientific Perspectives on Vedic Philosophy: A Treatise on the Unity of Scientific and Spiritual Knowledge. iUniverse.

Hooper, D. (2018). At the Edge of Time: Exploring the Mysteries of Our Universe's First Seconds. Princeton University Press.

Hume, R. E. (1921). The Thirteen Principal Upanishads. Oxford University Press.

Indich, W.M. (1980). Consciousness in Advaita Vedanta. Motilal Banarsidass.

Isayeva, N. (1993). Shankara and Indian Philosophy. State University of New York Press.

James, W. (1902). The Varieties of Religious Experience: A Study in Human Nature. Longmans, Green & Co.

Jitatmananda, S. (1986). Modern Physics and Vedanta. Bharatiya Vidya Bhavan.

Jitatmananda, S. (1993). Holistic Science and Vedanta. Bharatiya Vidya Bhavan.

Jonas, H. (1984). The Imperative of Responsibility: In Search of an Ethics for the Technological Age. University of Chicago Press.

Kafatos, M., & Nadeau, R. (1990). The Conscious Universe: Part and Whole in Modern Physical Theory. Springer-Verlag.

Kafatos, M., & Nadeau, R. (2000). The Conscious Universe: Parts and Wholes in Physical Reality. Springer.

Kak, S. (2000). The Astronomical Code of the Rgveda. Munshiram Manoharlal Publishers.

Kak, S. (2008). The Nature of Physical Reality. Mt. Meru Publishing.

Kak, S. (2021). Consciousness in Indian Philosophy: The Advaita Doctrine of 'Awareness Only'. Routledge.

Kaku, M. (2021). The God Equation: The Quest for a Theory of Everything. Doubleday.

Kastrup, B. (2019). The Idea of the World: A Multi-Disciplinary Argument for the Mental Nature of Reality. Iff Books.

Kauffman, S. A. (1995). At Home in the Universe: The Search for Laws of Self- Organization and Complexity. Oxford University Press.

Kineman, J. J. (2017). A causal framework for integrating contemporary and Vedic holism. Progress in biophysics and molecular biology, 131, 402-423.

Kirshner, R. P. (2002). The Extravagant Universe: Exploding Stars, Dark Energy, and the Accelerating Cosmos. Princeton University Press.

Klostermaier, K. K. (2000). Hindu Writings: A Short Introduction to the Major Sources. Oneworld Publications.

Koch, C. (2012). Consciousness: Confessions of a Romantic Reductionist. MIT Press. 7. Maharaj, N. (1973). I Am That: Talks with Sri Nisargadatta Maharaj. Acorn Press.

Kragh, H. (1996). Cosmology and Controversy: The Historical Development of Two Theories of the Universe. Princeton University Press.

Krauss, L. M. (2012). A Universe from Nothing: Why There Is Something Rather than Nothing. Free Press.

Krishnananda, S. (1994). The Philosophy of Life. The Divine Life Society.

Kuhn, R. L. (2024). A landscape of consciousness: Toward a taxonomy of explanations and implications. Progress in Biophysics and Molecular Biology.

Lanza, R., & Berman, B. (2009). Biocentrism: How Life and Consciousness are the Keys to Understanding the True Nature of the Universe. BenBella Books.

Lao Tzu. (1963). Tao Te Ching. (D. C. Lau, Trans.). Penguin Classics.

Laszlo, E. (2004). Science and the Akashic Field: An Integral Theory of Everything. Inner Traditions.

Leslie, J. (1989). Universes. Routledge.

Liddle, A. (2015). An Introduction to Modern Cosmology. Wiley. Ryden, B. (2016). Introduction to Cosmology. Cambridge University Press.

Linde, A. (1986). Eternally Existing Self-Reproducing Chaotic Inflationary Universe. Physics Letters B, 175(4), 395-400.

Lovelock, J. (2000). Gaia: A New Look at Life on Earth. Oxford University Press.

Loy, D. (1988). Nonduality: A Study in Comparative Philosophy. Yale University Press.

Malhotra, R. (2011). Being Different: An Indian Challenge to Western Universalism. HarperCollins.

Martchev, M. (2015). The Modern Discourse on Consciousness: Science Meets Vedantic Wisdom. 実践女子大学 CLEIP ジャーナル, 1, 87-103.

Menon, S. (2006). Advaita Vedanta. Internet Encyclopedia of Philosophy.

Menon, S. (2014). Brain, Self and Consciousness: Explaining the Conspiracy of Experience. Springer India.

Mohrhoff, U. (2007). The Quantum World, the Mind, and the Cookie Cutter Paradigm. Antimatters, 1(1), 55-90.

Mohrhoff, U. (2014). Manifesting the Quantum World. Philosophia Naturalis, 51(2), 267-281.

Muller, M. (1919). The Six Systems of Indian Philosophy. Longmans, Green and Co.

Muller, R. A. (2016). Now: The Physics of Time. W. W. Norton & Company.

Musser, G. (2018). Spooky Action at a Distance: The Phenomenon That Reimagines Space and Time--and What It Means for Black Holes, the Big Bang, and Theories of Everything. Scientific American / Farrar, Straus and Giroux.

Naess, A. (1989). Ecology, Community and Lifestyle: Outline of an Ecosophy. Cambridge University Press.

Nagel, T. (1974). What Is It Like to Be a Bat? The Philosophical Review, 83(4), 435-450.

Narlikar, J. V. (1993). Introduction to Cosmology. Cambridge University Press.

Needleman, J. (2016). Time and the Soul: Where Has All the Meaningful Time Gone - And Can We Get It Back? Berrett-Koehler Publishers.

Oberhummer, H., Csoto, A., & Schlattl, H. (2000). Stellar Production Rates of Carbon and Its Abundance in the Universe. Science, 289(5476), 88-90.

Olivelle, P. (1998). The Early Upanishads: Annotated Text and Translation. Oxford University Press.

Palgrave.Gupta, B. (1998). The Disinterested Witness: A Fragment of Advaita Vedanta Phenomenology. Northwestern University Press.

Panda, N. C. (1991). Maya in Physics. Motilal Banarsidass Publishers.

Panikkar, R. (2001). The Vedic Experience: Mantramanjari. Motilal Banarsidass.

Panikkar, R. (2010). The Rhythm of Being: The Gifford Lectures. Orbis Books.

Peebles, P. J. E. (1993). Principles of Physical Cosmology. Princeton University Press.

Penrose, R. (1989). The Emperor's New Mind: Concerning Computers, Minds, and the Laws of Physics. Oxford University Press.

Penrose, R. (2010). Cycles of Time: An Extraordinary New View of the Universe. The Bodley Head.

Popper, K. (1959). The Logic of Scientific Discovery. Hutchinson & Co.

Potter, K. H. (Ed.). (1981). Encyclopedia of Indian Philosophies, Vol. 3: Advaita Vedanta up to Samkara and His Pupils. Princeton University Press.

Price, H. (1996). Time's Arrow and Archimedes' Point: New Directions for the Physics of Time. Oxford University Press.

Prigogine, I. (1997). The End of Certainty: Time, Chaos, and the New Laws of Nature. The Free Press.

Pruett, D. (2017). Reason and Wonder: A Copernican Revolution in Science and Spirit. Praeger.

Radhakrishnan, S. (1923). Indian Philosophy. Oxford University Press.

Radhakrishnan, S. (1953). The Principal Upanishads. Harper.

Radhakrishnan, S. (1960). The Brahma Sutra: The Philosophy of Spiritual Life. George Allen & Unwin Ltd.

Radin, D. (2006). Entangled Minds: Extrasensory Experiences in a Quantum Reality. Paraview Pocket Books.

Radin, D., Michel, L., Galdamez, K., Wendland, P., Rickenbach, R., & Delorme, A. (2012). Consciousness and the double-slit interference pattern: Six experiments. Physics Essays, 25(2), 157-171.

Raju, P.T. (1971). The Philosophical Traditions of India. University of Pittsburgh Press.

Ram-Prasad, C. (2001). Knowledge and Liberation in Classical Indian Thought.

Rees, M. (1999). Just Six Numbers: The Deep Forces That Shape the Universe. Basic Books.

Rees, M. (2003). Our Final Hour: A Scientist's Warning. Basic Books. 4. Radhakrishnan, S. (1926). Hindu View of Life. Allen & Unwin.

Reichenbach, H. (1956). The Direction of Time. University of California Press.

Revonsuo, A. (2009). Consciousness: The Science of Subjectivity. Psychology Press.

Ricard, M., & Thuan, T. X. (2001). The Quantum and the Lotus: A Journey to the Frontiers Where Science and Buddhism Meet. Crown.

Rieper, E., Anders, J., & Vedral, V. (2011). Quantum entanglement between the electron clouds of nucleic acids in DNA. arXiv preprint arXiv:1006.4053.

Rioux, S. (1997). Quantum physics and Vedanta: a perspective from Bernard d'Espagnat's scientific realism. Zygon, 32(2), 235-248.

Rolston III, H. (1988). Environmental Ethics: Duties to and Values in the Natural World. Temple University Press.Vivekananda, S. (1896). Karma Yoga. Longmans, Green, and Co.

Rosenblum, B., & Kuttner, F. (2011). Quantum Enigma: Physics Encounters Consciousness. Oxford University Press.

Rovelli, C. (2016). Reality Is Not What It Seems: The Journey to Quantum Gravity. Riverhead Books.

Rovelli, C. (2018). The Order of Time. Riverhead Books.

Sagan, C. (1985). Contact. Simon & Schuster.

Sagan, C. (1994). Pale Blue Dot: A Vision of the Human Future in Space. Random House.

Sankara. (1949). Brahma Sutra Bhasya. (S. Gambhirananda, Trans.). Advaita Ashrama.

Sarma, D. (2011). Classical Indian Philosophy: A Reader. Columbia University Press.

Satchidananda Murty, K. (1959). Revelation and Reason in Advaita Vedanta. Columbia University Press.

Schrödinger, E. (1944). What is Life? The Physical Aspect of the Living Cell. Cambridge University Press.

Searle, J. R. (1997). The Mystery of Consciousness. New York Review of Books.

Sharma, A. (1993). The Experiential Dimension of Advaita Vedanta. Motilal Banarsidass Publishers.

Sharma, A. (1995). The Philosophy of Religion and Advaita Vedanta: A Comparative Study in Religion and Reason. Pennsylvania State University Press.

Sharma, C. (1960). A Critical Survey of Indian Philosophy. Motilal Banarsidass.

Shear, J. (ed.) (1997). Explaining Consciousness: The Hard Problem. MIT Press.

Sheldrake, R. (2012). The Science Delusion: Freeing the Spirit of Enquiry. Coronet

Sidorova-Biryukova, A. (2020). Theoretical Physics and Indian Philosophy: Conceptual Coherence. arXiv preprint arXiv:2004.02150.

Silk, J. (2009). Horizons of Cosmology. Templeton Press.

Singh, J. (1961). Great Ideas and Theories of Modern Cosmology. Dover Publications.

Singh, S. (2004). Big Bang: The Origin of the Universe. Fourth Estate.

Skolimowski, H. (1994). The Participatory Mind: A New Theory of Knowledge and of the Universe. Penguin Books India.

Smolin, L. (1997). The Life of the Cosmos. Oxford University Press.

Smolin, L. (2006). The Trouble with Physics: The Rise of String Theory, The Fall of a Science, and What Comes Next. Houghton Mifflin.

Smolin, L. (2013). Time Reborn: From the Crisis in Physics to the Future of the Universe. Houghton Mifflin Harcourt.

Stapp, H. P. (2007). Mindful Universe: Quantum Mechanics and the Participating Observer. Springer.

Steinhardt, P. J., & Turok, N. (2007). Endless Universe: Beyond the Big Bang. Doubleday.

Susskind, L. (2005). The Cosmic Landscape: String Theory and the Illusion of Intelligent Design. Little, Brown and Company.

Swami Jitatmananda. (1991). Modern Physics and Vedanta. Bhavan's Book University.

Swimme, B., & Berry, T. (1994). The Universe Story: From the Primordial Flaring Forth to the Ecozoic Era. HarperOne.

Talbot, M. (1991). The Holographic Universe: The Revolutionary Theory of Reality. HarperCollins.

Tegmark, M. (2014). Our Mathematical Universe: My Quest for the Ultimate Nature of Reality. Alfred A. Knopf.

Teilhard de Chardin, P. (1959). The

Theise, N. D., & Kafatos, M. C. (2016). Fundamental awareness: A framework for integrating science, philosophy and metaphysics. Communicative & Integrative Biology, 9(3), e1155010.

Thompson, E. (2014). Waking, Dreaming, Being: Self and Consciousness in Neuroscience, Meditation, and Philosophy. Columbia University Press.

Tononi, G. (2012). Integrated information theory of consciousness: an updated account. Archives Italiennes de Biologie, 150(2-3), 56-90.

Tyson, N. D., & Goldsmith, D. (2004). Origins: Fourteen Billion Years of Cosmic Evolution. W. W. Norton & Company.

Van Strien, M. (2020). Bohm's theory of quantum mechanics and the notion of classicality. Studies in History and Philosophy of Science Part B: Studies in History and Philosophy of Modern Physics, 71, 72-86.

Velmans, M. (2009). Understanding Consciousness. Routledge.

Vimal, R. L. P. (2009). Meanings attributed to the term 'consciousness': an overview. Journal of Consciousness Studies, 16(5), 9-27.

Vimal, R. L. P. (2009). Meanings attributed to the term 'consciousness': an overview. Journal of Consciousness Studies, 16(5), 9-27.

Vivekananda, S. (1896). Raja Yoga. Longmans, Green, and Co.

Vivekananda, S. (1907). Jnana-Yoga. Ramakrishna-Vivekananda Center.

Wallace, B. A. (2007). Contemplative Science: Where Buddhism and Neuroscience Converge. Columbia University Press.

Weinberg, S. (1977). The First Three Minutes: A Modern View of the Origin of the Universe. Basic Books.

Weinberg, S. (1987). Anthropic Bound on the Cosmological Constant. Physical Review Letters, 59(22), 2607.

Weinberg, S. (1992). Dreams of a Final Theory: The Scientist's Search for the Ultimate Laws of Nature. Pantheon Books.

Wheeler, J. A. (1977). Genesis and Observership. In R. E. Butts & J. Hintikka (Eds.), Foundational Problems in the Special Sciences. Reidel.

Wheeler, J. A., & Zurek, W. H. (Eds.). (1983). Quantum Theory and Measurement. Princeton University Press.

Whitrow, G. J. (1988). Time in History: Views of Time from Prehistory to the Present Day. Oxford University Press.

Wilber, K. (1997). The Eye of Spirit: An Integral Vision for a World Gone Slightly Mad. Shambhala.

Wilber, K. (2000). A Theory of Everything: An Integral Vision for Business, Politics, Science, and Spirituality. Shambhala.

Wilczek, F. (2015). A Beautiful Question: Finding Nature's Deep Design. Penguin Press.

Wilson, E. O. (1998). Consilience: The Unity of Knowledge. Knopf.

Wolf, F. A. (1981). Taking the Quantum Leap: The New Physics for Nonscientists. Harper & Row.

Zeman, A. (2002). Consciousness: A User's Guide. Yale University Press.

Zukav, G. (1979). The Dancing Wu Li Masters: An Overview of the New Physics. William Morrow and Company.

Zukav, G. (1979). The Dancing Wu Li Masters: An Overview of the New Physics. William Morrow and Company.

Index

A

B

E

Education and Consciousness, 95

Edwin Hubble, 8

Einstein's Relativity, 37

Emanation vs. Creation, 32

Emergent Time, 41

Empirical reality, 132

Entropy and Cosmic Cycles, 106

Entropy and Environmental Ethics, 159

Environmental Ethics, 122, 164, 242

Environmental Philosophy, 183

Epistemology, 182

Eternal Cycles, 105

eternal now, 40

Eternity, 27, 35, 45, 50, 51, 53, 54, 55, 56, 231, 232

Eternity and Immortality, 45

Ethical Implications, 4

Ethics, 55, 94, 157, 164, 170, 171, 182, 236

Ethics and Consciousness, 94

Evolution and Ethics, 159

Evolutionary Ethics, 172

Evolving Consciousness, 187

Evolving Philosophical Frameworks, 138

Expanded States of Consciousness, 154

Expanded Time Scales, 162

Expanding Universe, 8

Expansion of the Universe, 11

Extraterrestrial Life and Expanding Moral Circles, 160

F

Fate of the Universe, 15, 108, 215

Formation of Elements, 11

Formation of Planetary Systems, 102

Four States of Consciousness, 146

Free Will and Determinism, 43, 72, 94, 134, 148

Fundamental Uncertainty, 147

Future of Unification, 137

G

General Challenges to Unification, 135

General Relativity, 37, 128

Georges Lemaître, 9

Global Workspace Theory, 84

Gödel's Incompleteness Theorems, 144

H

Heisenberg Uncertainty Principle, 142

Heisenberg's Uncertainty Principle, 62

Holistic Nature of Reality, 227

Holistic Worldviews, 33, 137

Human Lifespan, 55

I

ignorance (avidya), 27, 132

Illusory reality, 68, 132

Implications of a Unified Cosmic Vision, 182

ineffability of Brahman, 147

Ineffability of Brahman, 145

Inflation Theory, 13

Integrating Perspectives, 96, 110, 122